THE STATE OF THE EARTH

THE STATE OF THE EARTH
Edited by Joni Seager

A PLUTO PROJECT

UNWIN

HYMAN

LONDON   SYDNEY   WELLINGTON

First published 1990 by
Unwin Hyman Limited,
15–17 Broadwick Street, London W1V 1FP

ISBN: 0 04 440693 2 HBK
       0 04 440692 4 PBK

Edited and coordinated by Anne Benewick

Design by Grundy & Northedge, London

Artwork by James Mills-Hicks, Andrea Fairbrass,
Jacqueline Land and Jeanne Radford for
Swanston Graphics Ltd, Derby

Cover design by Ian Cockburn
Cover photo of rainforest in the Solomon Islands
by Oxford Scientific Films Ltd

Printed and bound in Hong Kong by
Mandarin Offset International (HK) Ltd

1   2   3   4   5   6   7   8   9   10

This book is printed on paper made from pulp from
Finland and Sweden. It comes from managed
forest, where the rate of replanting means that
timber stocks are rising.

British Library Cataloguing in Publication Data
Seager, Joni
  The state of the earth atlas.
  1. Environment. Effects of man
  I. Title
  333.72

  ISBN 0-04-440693-2
  ISBN 0-04-440692-4 pbk

CONTENTS

EDITOR

JONI SEAGER is a Canadian geographer, currently living in the United States. She is a lecturer at the Massachusetts Institute of Technology and a member of the New Words feminist book collective. She is the co-author of *Women in the World: An International Atlas* and author of *Earth Follies: Making Feminist Sense of the Global Environment* (forthcoming).

CONTRIBUTORS

MARINA ALBERTI works on issues of hazardous waste management and environmental policy. She is a member of the Cooperativa Ecologia in Milan, where she participates actively in the environmental movement.

SIMON ALBRECHT is currently running the Data Support for Education service for the Nature Conservancy Council, Great Britain. He organised a seminar on Teaching Conservation in 1986 and has written two educational resource packs, *Pollution Facts* and *Conservation Facts* (1989).

MIKE BIRKIN has a long involvement with Friends of the Earth, UK and is currently their Regional Development Officer in London. Formerly an environmental consultant, he is co-author of *C is for Chemicals: Chemical Hazards and How to Avoid Them* (1989).

ANDY CRUMP has worked in various parts of the world as an ecologist, biological pest control specialist and writer on all aspects of environment and development. Formerly with the Panos Institute in London, he is currently a freelance writer/researcher in West Germany.

LESLEY DOWNING is a specialist in rural development who has worked in Kenya and the Philippines for seven years, administering community-based development programmes and including projects on small-scale rural water supply.

THOMAS E. DOWNING is a geographer who has conducted research on environmental issues, natural hazards, drought, famine and food policy in Africa, the USA and Mexico. He has spent four years in Kenya formulating district environmental policies.

NIGEL DUDLEY is an ecologist, specializing in pollution issues and the impact of Western institutions on the environment of the South. He works particularly with Earth Resources Research in London and the Soil Association in Bristol.

DOMINIC GOLDING is a Senior Research Associate at the Center for Technology, Environment and Development (CENTED), Clark University, Massachusetts. He is a geographer with research interests in the social aspects of risk assessment and hazard management.

LISA GREBER is a physicist and an activist who has worked on solar cell materials and citizens' control of nuclear power. She is currently acting director of Science for the People in Cambridge, Massachusetts.

CLAIRE HOLMAN is an independent environmental researcher working mainly on air pollution issues. She has worked as Information Officer for Friends of the Earth, UK and is currently adviser to both Friends of the Earth, UK and the Worldwide Fund for Nature, UK.

CONSTANCE E. HUNT is a biologist working for the US Army Corps of Engineers. She was previously a Research Associate at the National Wildlife Federation (US) and is the author of *Down by the River* (1988).

CHRISTINE LANCASTER is currently working as an Education Campaigner for the Parents Against Tobacco campaign (UK). Previously, she was a Research Associate on the Friends of the Earth Tropical Rain Forest Campaign.

JEAN MACRAE is an independent researcher in Boston, Massachusetts with an interest in public health. She is a founding member of the New Words Feminist Bookstore Collective.

JAN MCHARRY is currently Information Officer for Friends of the Earth, UK. Trained as an environmental resource planner, she has had a long involvement with grassroots environmental groups in the UK working on a range of environmental issues.

MICHAEL RENNER is a Senior Researcher at the Worldwatch Institute in Washington DC. He is working on environmental security issues.

MICHAEL SHANABROOK is a research assistant at the Congressional Budget Office in Washington DC. His research interests include the impact of global warming and international environmental treaty negotiations.

DAN SMITH is Associate Director of the Transnational Institute, Amsterdam and the author of several books on armament and disarmament. He is co-author with Michael Kidron of another Pluto Project, *The War Atlas* (1983), to appear in a second edition in 1991 as *The New Atlas of War and Peace*.

MICHAEL STEINITZ has taught urban and cultural geography and researched topics relating to past human environmental interactions. He is presently working in Massachusetts for the protection of historic built environments and sites.

ACKNOWLEDGEMENTS

For their assistance in providing information for individual maps, we would like to thank the following:

Hayward R. Alker, Jr., Massachusetts Institute of Technology, Cambridge, Mass.; Carl Bartone, World Bank, Washington DC; Margaret Bluman, London; Steven Broad, World Conservation Monitoring Centre, Cambridge, UK; Marva Coates, Food and Agriculture Organization, Washington DC; Peter Cebon, Massachusetts Institute of Technology, Cambridge, Mass.; Mark Davis, London; Stephen Davis, World Conservation Monitoring Centre, Surrey, UK; Cynthia Enloe, Clark University, Worcester, Mass.; Environmental Project on Central America (EPOCA); Joe Farman, British Antarctic Survey, Cambridge, UK; Paula Ferguson, Massachusetts Institute of Technology, Cambridge, Mass.; Hilary French, Worldwatch Institute, Washington DC; John Gille, National Center for Atmospheric Research, Colorado; Rupert Hastings, Royal Botanic Gardens, Surrey, UK; Norbert Henninger, World Resources Institute, Washington DC; Richard Hosier, University of Pennsylvania; Carol Inskipp, Worldwide Fund for Nature, Surrey, UK; Jodi Jacobson, Worldwatch Institute, Washington DC; Judy Johnson, Environmental Protection Agency, Washington DC; Peter Kearns, Department of the Environment, UK; Victoria Lane, Somerville, Mass.; Nick Lenssen, Worldwatch Institute, Washington DC; Christine Leon, World Conservation Monitoring Centre, Surrey, UK; Robert Mason, Temple University, Philadelphia; Alison McCusker, International Board for Plant Genetic Resources, Rome; Christine McGowan, Advisory Committee on Genetic Manipulation, UK; Arthur McKenzie, Tanker Advisory Center, New York City; Maldives Services Ltd, London; Norman Myers; Nature Conservancy Council, Data Support for Education Service, Peterborough, UK; Sam Ozolins, World Health Organization, Geneva; Steve Parcels, Natural Resources Defense Council, New York City; Bonnie Ram, Washington DC; Scott Saleska, Institute for Energy and Environmental Research, Takoma Park, Maryland; Cindy Pollock Shea, Worldwatch Institute, Washington DC; Seth Shulman, Science for the People, Cambridge, Mass.; Teresa Squires, American Cancer Society; Koy Thompson, Friends of the Earth, UK; World Action on Recycling Materials for Energy from Rubbish (WARMER), Tunbridge Wells, UK; Worldwide Fund for Nature, Data Support for Education Service, Surrey, UK; Worldwide Fund for Nature International, Switzerland.

INTRODUCTION

Concern is the catalyst that shapes this atlas. We share with many the belief that habitat earth, in the late 20th century, is in growing trouble. This is not really news; as the media, advertisers and even industry take up the 'green' cause, this unsettling assessment can scarcely have escaped the attention of many.

And yet much information about environmental matters remains confined to the privileged domain of specialists. Our purpose in undertaking this atlas is to translate specialist information into a form that can be more widely appreciated. As environmental consciousness grows to a popular groundswell, we need to be able to make sense of the mass of facts bombarding us from our newspapers and television screens. The need for a coherent overview of the state of the earth has never been greater, particularly so for information presented in an accessible and engaging manner.

In this atlas, we use the fact of environmental crisis not as an endpoint, but as the place from which to begin further enquiry. Not all parts of the global environment and not all places in the world are in the same state of distress. We ask where threats exist to a greater or lesser degree, and why they are so distributed. Most of the deterioration in our environment is the direct or indirect consequence of particular human arrangements of economic and political power, superimposed on vulnerable ecological systems. Not all economies nor all cultures, however, impose the same burdens on the environmental equilibrium. This atlas seeks to identify where such threats originate and why, indeed, they do so.

We encompass the state of the earth as broadly as possible. Thus, we offer information on issues that may already be familiar to many, such as ozone depletion, tropical rainforest destruction, and the plague of oil spills. But we also extend our reach to include topics that have received less popular attention, such as tourism, car culture, wetlands loss, and the international trade in tropical timber and in wildlife.

A primary truth of environmentalism is that everything is connected. The web of life is seamless, and the consequences of disruption to one part of the ecosystem ripple throughout the whole: soil erosion in the Himalayas contributes to massive flooding in Bangladesh, the deforestation of the Amazon may alter the atmospheric balance over the whole globe, and chemicals and gases produced in the richer industrial countries are destroying the ozone layer that protects everyone, rich and poor alike.

This interconnectedness makes it an excercise of frustration to arrange environmental topics in a clearly demarcated, apparently logical sequence. It should be understood that both sections and subject sequence in this atlas are largely artificial. The topics we have chosen are essentially part of a synergistic whole, wherein as much information can be gleaned from drawing comparisons and contrasts between maps as from studying them individually. The use of maps does, at least, facilitate such cross-referencing; continuities and contrasts become visible patterns on the page; threads of similarity and difference weave their way through the book.

It is a startling fact, which this atlas makes clear, that much environmental degradation transcends traditional economic and political boundaries. Environmental problems are much the same in centrally-controlled Communist states and 'free market' capitalist ones, in people's republics and diehard monarchies. The residents of Beijing and Paris suffer much the same levels of sulphur dioxide poisoning in the air they breathe; the impact of soil erosion is much the same whether in Jamaica or the USA; per head of population, Saudi Arabia, the UK and the USSR emit similar levels of carbon dioxide into the world's atmosphere; both the Aral Sea and the North Sea are dying because of the misplaced priorities of governments and industries, East and West.

When trying to make sense of environmental issues, the conventional way of conceptualizing world divisions by political boundaries seems inadequate. Dividing the world into communist and capitalist states, or free market and state-controlled economies, is of limited value in explaining the environmental state of the earth. More useful connections can be made between those countries that are fossil-fuel dependent and those that are hydro-dependent; or between those with and without tropical rainforest; or between those that produce CFCs and those that do not. By creating new links and categories of explanation in this atlas, we enable alliances to be reconceptualized: the USA and USSR have more in common than not, and the biggest blocs of environmental alliance are among rich and industrial nations and among poor and non-industrial ones.

New alliances based on environmental affinities and commonalities are already forming and will shape environmental politics into the next century. In 1989, the Maldives government hosted the first 'Small States Conference on Sea Level Rise', bringing together oceanic states that will be the first affected by global warming; in the late 1980s, a number of caucuses of poorer countries, including meetings of the Non-Aligned States, the Economic Community of West African States, and the Organization of African Unity, passed resolutions calling on industrialized countries to halt toxic waste dumping; and the '30 percent Club' has brought together those countries of Western Europe committed to cooperate in reducing sulphur dioxide emissions, the major cause of acid rain.

Certainly the weight of evidence underscores the responsibility of rich states for many of the global environmental problems we now face: the loss of animal, bird and plant species, fossil fuel pollution, the production and use of ozone-depleting chemicals, the ocean pollution that comes with oil dependency, the sulphur dioxide emissions that destroy forests and lakes, the stripping of tropical forests for timber and cattle-ranch farming - to name but a few. The catalogue of problems caused by the voracious resource appetites of the rich is almost overwhelming and the by-products are fouling our environment. The rich industrial world is stripping the earth of resources and in the course of so doing is generating enormous quantities of effluent and pollution that burden global ecosystems.

Rich countries continue to enhance their wealth by expanding their resource reach beyond their own borders while frequently exporting their problems. Having stripped their own forests, they are turning to timber-plundering in former colonies; choking on a glut of municipal, household and industrial garbage, they seek to ship away their waste to small Pacific islands or debt-strapped countries in Africa. But this is a small planet: there are limited resource frontiers to exploit - and there is no 'away'. There is no safe place to store carcinogenic chemicals, nuclear waste or the toxic by-products of industry. There is no safe way of dispersing CFCs or many of the other harmful gases being vented into the atmosphere. There are limits to global tolerance.

The global scale of this atlas makes powerful generalizations about the responsibilities of the rich world. It is less able to expose the local examples of environmental degradation that are all around us. Yet the destruction of wildlife habitat, the pollution of waterways and fuelwood scarcity – to name but a few issues – are often at their most serious at a regional or site-specific level.

On this smaller scale, it is important to recognize that the role of individuals in influencing environmental fates is mediated by, among other factors, gender, class, ethnicity and religion. Men and women, for example, have different relationships to the large institutions – governments, militaries, industrial corporations – that hold the balance of power on environmental issues. Similarly, the implications and experience of environmental decay are often different for men and women, rich and poor, elites and disenfranchized. Within a country, region or city, environmental degradation will be experienced differently, and more direly, by poor people living in slums with limited access to safe water, food and health care, than wealthier people insulated from environmental problems by their privilege.

The maps in this atlas inevitably focus on the physical form of environmental problems. With this in mind, in the Commentary to the maps (see page 99*ff*) we have tried wherever possible to discuss social, political and gender issues.

There are important environmental issues that, for one reason or another, do not appear in this atlas. Mining, perhaps because it includes a wide range of activities, defied our concept of global mapping even though it is a universally destructive enterprise, causing land degradation, pollution from tailings, chemical leaching, deforestation and landscape scarring. Although car culture and its concomitant air pollution are both included, the increasing density of road networks is omitted, for lack of appropriate global statistics. For similar reasons we do not map the spread of industrialization, the displacement of subsistence agriculture by cash cropping and luxury foods, the increasing numbers of environmental refugees, or grassroots mobilization to confront environmental problems.

Global information-gathering technology is proliferating. At the same time, access to and control of information is becoming more privatized. Full disclosure of environmental problems does not necessarily serve the purposes of the industries and governments that gather most environmental data. Without the untiring work of independent environmental groups such as the World Resources Institute, Greenpeace, Friends of the Earth and other international, national and community-based groups, we would know even less than we do about the state of the earth. We would like to think that this atlas will support and contribute to their efforts. An informed citizenry is the world's best hope.

The creation of this atlas drew on the goodwill and good sense of many dozens of individuals. I extend elsewhere our collective thanks to those who provided assistance with particular maps. I wish to single out for appreciation here a few brave souls who gave encouragement, criticism and invaluable support through every stage of this project. My greatest debt is to Anne Benewick, whose keen eye and wit shaped this project from its very beginning, and who shepherded it through all the editorial and production stages from mere doodles on paper to finished artwork and text. David Pratt made a valuable contribution in the formative stages of the project and Grundy & Northedge added their fine and distinctive design style. Swanston Graphics proved stellar, true to form, and an editor could not ask for a more competent and pleasant production and support team. Two project advisers, John Baines in the UK and Jeanne Kasperson of the Center for Technology, Environment & Development at Clark University, assisted in the early conceptualization of the project and, in particular, helped to assemble our research team.

Cynthia Enloe's unwavering support sustained me throughout the year-and-a-half that this atlas took over my life, and I have drawn heavily on her analytical acuity to steer a course through the intellectual maze of trying to make sense of the state of the earth. My colleagues in the New Words collective – Gilda Bruckman, Mary Lowry, Doris Reisig, Kate Rushin and Laura Zimmerman – tolerated my more-than-usual eccentricities and accommodated my absences with good humour.

The compilation of the research for this atlas was a collaborative effort in which I was aided by an able team of contributors who each provided not only specialist knowledge and timely insight, but a sense of camaraderie and shared purpose. The strength of this atlas rests on the firm research foundation which they built, map by map. For any errors that slipped into the presentation of their research, or for the liberties that I may have taken in its interpretation, I take sole responsibility.

Joni Seager
Somerville, Massachusetts
April 1990

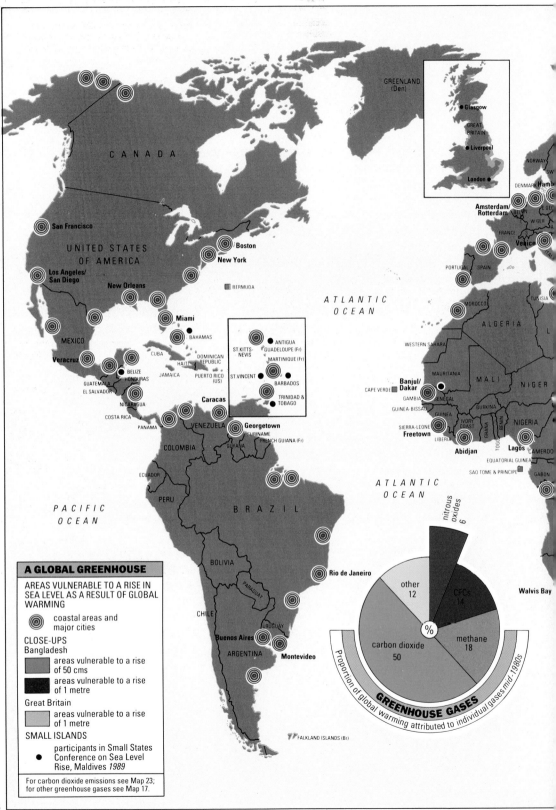

GREENLAND
(Den)

• Glasgow

GREAT
BRITAIN

• Liverpool

London •

NORWAY

DENMARK **Hamb**

**Amsterdam/
Rotterdam**

BEL N
W GER
NER

FRANCE

Veni ce

PORTUGAL SPAIN

TUNISIA

MOROCCO

ALGERIA

WESTERN SAHARA

MAURITANIA

MALI

NIGER

**Banjul/
Dakar**

CAPE VERDE

GAMBIA SENEGAL
GUINEA-BISSAU
GUINEA
BURKINA

IVORY
COAST

BENIN

NIGERIA

SIERRA LEONE
Freetown

LIBERIA

GHANA

TOGO

Lagos

CAMEROO

Abidjan

EQUATORIAL GUINEA
SAO TOME & PRINCIPE

GABON

CANADA

ATLANTIC
OCEAN

San Francisco

UNITED STATES
OF AMERICA

Boston

New York

**Los Angeles/
San Diego**

New Orleans

BERMUDA

Miami

BAHAMAS

MEXICO

CUBA

DOMINICAN
REPUBLIC

Veracruz

BELIZE
HONDURAS

HAITI

JAMAICA

PUERTO RICO
(US)

ST.KITTS-
NEVIS

ANTIGUA
GUADELOUPE (Fr)
MARTINIQUE (Fr)

ST.VINCENT

BARBADOS

GUATEMALA
EL SALVADOR

NICARAGUA

Caracas

TRINIDAD &
TOBAGO

COSTA RICA

PANAMA

VENEZUELA

Georgetown

SURINAME
GUYANA
FRENCH GUIANA (Fr)

ATLANTIC
OCEAN

COLOMBIA

ECUADOR

PERU

Walvis Bay

PACIFIC
OCEAN

B R A Z I L

BOLIVIA

PARAGUAY

Rio de Janeiro

CHILE

URUGUAY

Buenos Aires

ARGENTINA

Montevideo

FALKLAND ISLANDS (Br)

A GLOBAL GREENHOUSE

AREAS VULNERABLE TO A RISE IN
SEA LEVEL AS A RESULT OF GLOBAL
WARMING

◎ coastal areas and
major cities

CLOSE-UPS
Bangladesh

areas vulnerable to a rise
of 50 cms

areas vulnerable to a rise
of 1 metre

Great Britain

areas vulnerable to a rise
of 1 metre

SMALL ISLANDS

• participants in Small States
Conference on Sea Level
Rise, Maldives *1989*

For carbon dioxide emissions see Map 23;
for other greenhouse gases see Map 17.

GREENHOUSE GASES

Proportion of global warming attributed to individual gases mid-1980s

nitrous
oxides
6

CFCs
14

other
12

%

methane
18

carbon dioxide
50

16

Data compiled by Nigel Dudley   Main sources: National Center for Atmospheric Research/National Science Foundation; World Resources Institute (1988); Hansen (1988).

Present evidence of global warming is controversial, but the implications could be so severe that the threat must be taken seriously.

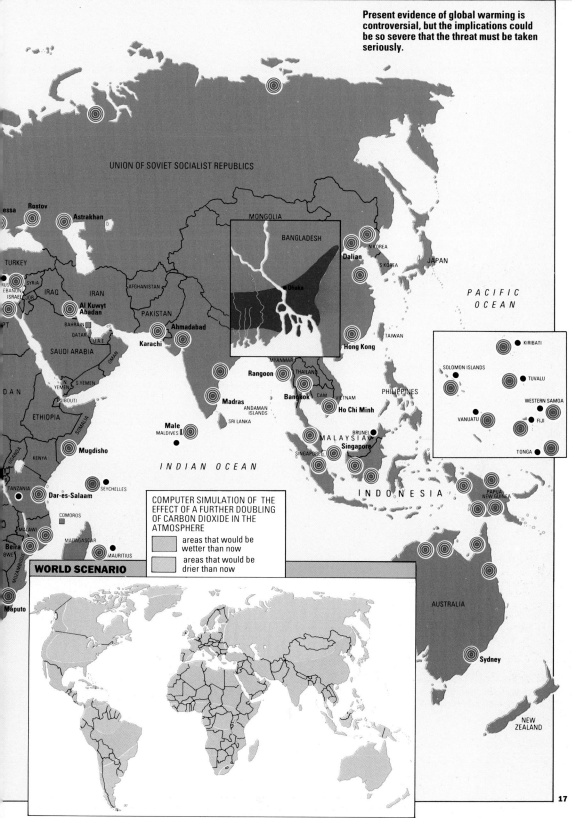

UNION OF SOVIET SOCIALIST REPUBLICS

MONGOLIA

BANGLADESH

Dhaka

essa
Rostov
Astrakhan
TURKEY
SYRIA
LEBANON
ISRAEL
IRAQ
IRAN
AFGHANISTAN
PAKISTAN
Al Kuwyt
Abadan
BAHRAIN
QATAR
U.A.E.
SAUDI ARABIA
OMAN
Karachi
Ahmadabad
N.KOREA
S.KOREA
JAPAN
Dalian
Hong Kong
TAIWAN

PACIFIC OCEAN

MYANMAR
THAILAND
Rangoon
PHILIPPINES
Bangkok
CAM
VIETNAM
Ho Chi Minh
Madras
ANDAMAN ISLANDS
SRI LANKA
Male
MALDIVES
MALAYSIA
BRUNEI
Singapore
SINGAPORE

INDIAN OCEAN

INDONESIA

N.YEMEN
S.YEMEN
DJIBOUTI
DAN
ETHIOPIA
SOMALIA
UGANDA
KENYA
Mugdisho
TANZANIA
SEYCHELLES
Dar-es-Salaam
COMOROS
MALAWI
MADAGASCAR
MAURITIUS
Beira
BWE
MOZAMBIQUE
Maputo

KIRIBATI
SOLOMON ISLANDS
TUVALU
WESTERN SAMOA
VANUATU
FIJI
TONGA

PAPUA NEW GUINEA

AUSTRALIA

Sydney

NEW ZEALAND

COMPUTER SIMULATION OF THE EFFECT OF A FURTHER DOUBLING OF CARBON DIOXIDE IN THE ATMOSPHERE

areas that would be wetter than now

areas that would be drier than now

WORLD SCENARIO

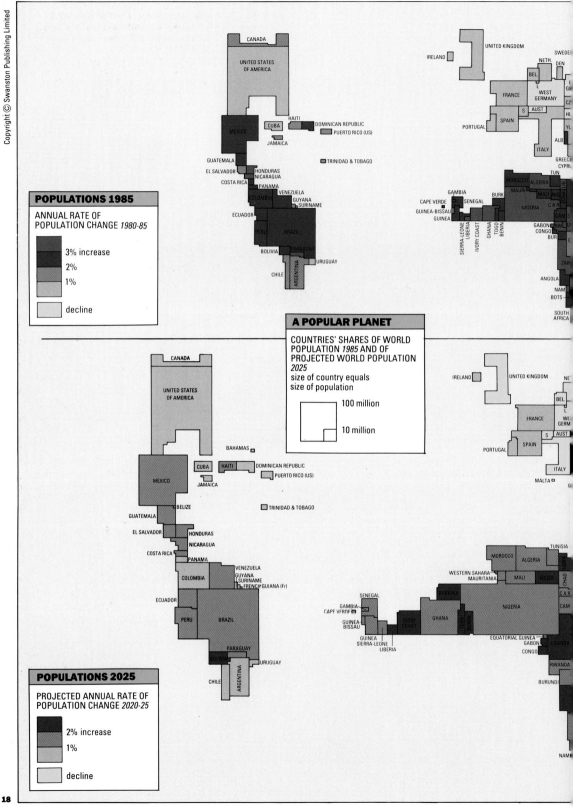

POPULATIONS 1985

ANNUAL RATE OF
POPULATION CHANGE *1980-85*

- 3% increase
- 2%
- 1%
- decline

A POPULAR PLANET

COUNTRIES' SHARES OF WORLD
POPULATION *1985* AND OF
PROJECTED WORLD POPULATION
2025
size of country equals
size of population

- 100 million
- 10 million

POPULATIONS 2025

PROJECTED ANNUAL RATE OF
POPULATION CHANGE *2020-25*

- 2% increase
- 1%
- decline

Data compiled by Michael Steinitz   Main sources: UN *World Population Prospects* (1988).

The average person in the 'developed' world consumes 10 times as much energy and one-and-a-half times as much food – and produces 16 times as much air pollution – as the average person in the 'developing' world. The deterioration of the global environment has less to do with population growth than with industrial economies and priorities.

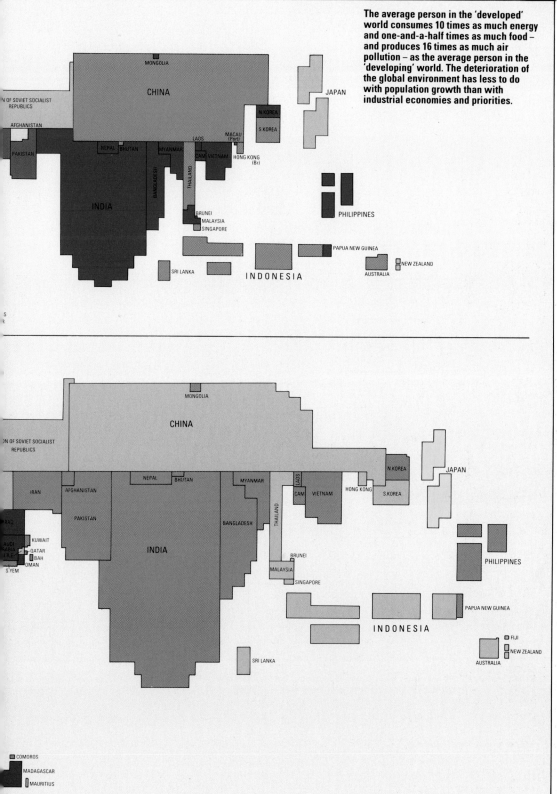

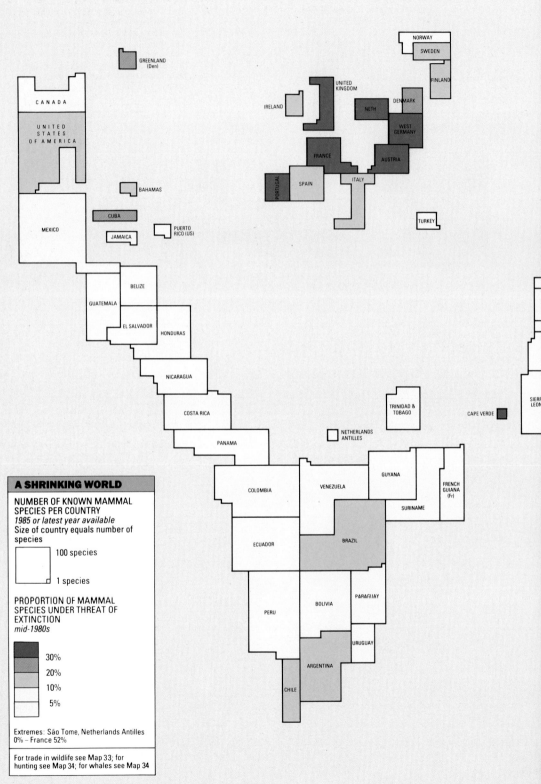

NORWAY

SWEDEN

FINLAND

GREENLAND
(Den)

CANADA

UNITED
STATES
OF AMERICA

UNITED
KINGDOM

IRELAND

NETH

DENMARK

WEST
GERMANY

FRANCE

AUSTRIA

PORTUGAL

SPAIN

ITALY

TURKEY

BAHAMAS

CUBA

JAMAICA

PUERTO
RICO (US)

MEXICO

BELIZE

GUATEMALA

EL SALVADOR

HONDURAS

NICARAGUA

COSTA RICA

PANAMA

TRINIDAD &
TOBAGO

CAPE VERDE

SIERRA
LEONE

NETHERLANDS
ANTILLES

GUYANA

COLOMBIA

VENEZUELA

FRENCH
GUIANA
(Fr)

SURINAME

ECUADOR

BRAZIL

PERU

BOLIVIA

PARAGUAY

URUGUAY

ARGENTINA

CHILE

A SHRINKING WORLD

**NUMBER OF KNOWN MAMMAL
SPECIES PER COUNTRY**
1985 or latest year available
Size of country equals number of
species

100 species

1 species

**PROPORTION OF MAMMAL
SPECIES UNDER THREAT OF
EXTINCTION**
mid-1980s

30%

20%

10%

5%

Extremes: São Tome, Netherlands Antilles
0% – France 52%

For trade in wildlife see Map 33; for
hunting see Map 34; for whales see Map 34

Data compiled by Andy Crump   Main sources: IUCN Monitoring Centre (1988); OECD (1989); World Resources Institute (1987, 1988)

By the close of this century, 100 animal species a day may be facing extinction.

EMPTY NESTS

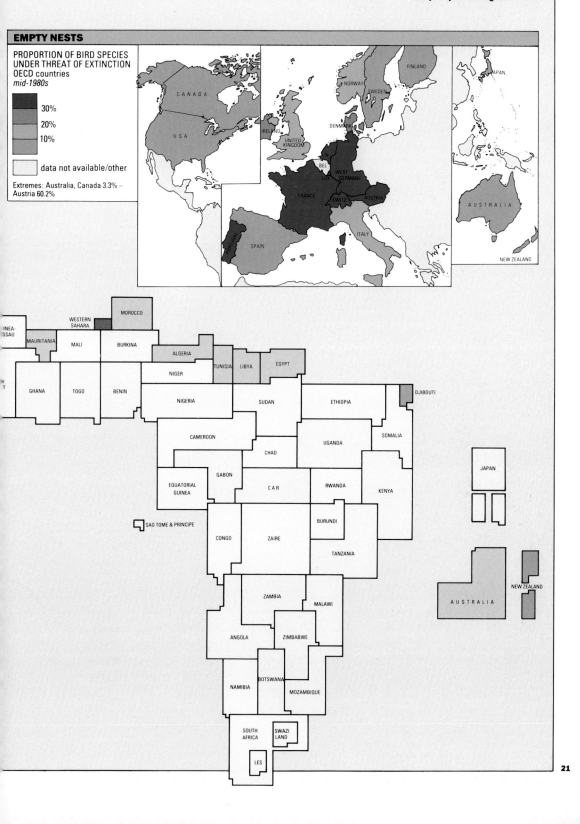

PROPORTION OF BIRD SPECIES
UNDER THREAT OF EXTINCTION
OECD countries
mid-1980s

- 30%
- 20%
- 10%

data not available/other

Extremes: Australia, Canada 3.3% –
Austria 60.2%

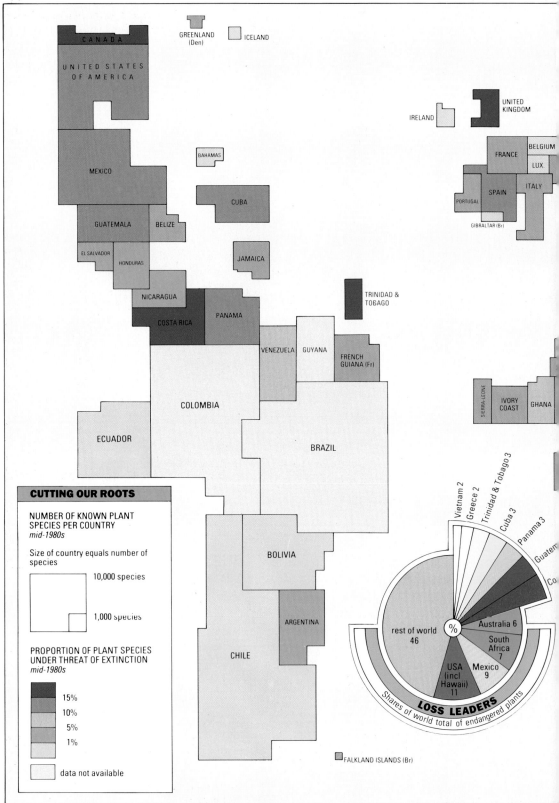

CANADA

GREENLAND
(Den)

ICELAND

UNITED STATES
OF AMERICA

IRELAND

UNITED
KINGDOM

BAHAMAS

MEXICO

FRANCE

BELGIUM

LUX.

PORTUGAL

SPAIN

ITALY

GIBRALTAR (Br)

CUBA

GUATEMALA

BELIZE

EL SALVADOR

HONDURAS

JAMAICA

NICARAGUA

TRINIDAD &
TOBAGO

COSTA RICA

PANAMA

VENEZUELA

GUYANA

FRENCH
GUIANA (Fr)

SIERRA LEONE

IVORY
COAST

GHANA

COLOMBIA

ECUADOR

BRAZIL

CUTTING OUR ROOTS

NUMBER OF KNOWN PLANT
SPECIES PER COUNTRY
mid-1980s

Size of country equals number of
species

10,000 species

1,000 species

PROPORTION OF PLANT SPECIES
UNDER THREAT OF EXTINCTION
mid-1980s

- 15%
- 10%
- 5%
- 1%

data not available

BOLIVIA

ARGENTINA

CHILE

Vietnam 2

Greece 2

Trinidad & Tobago 3

Cuba 3

Panama 3

Guatem...

Co...

rest of world
46

Australia 6

South
Africa
7

USA
(incl
Hawaii)
11

Mexico
9

%

LOSS LEADERS
Shares of world total of endangered plants

FALKLAND ISLANDS (Br)

Data compiled by Simon Albrecht   Main sources: Davis et al. (1979); World Resources Institute (1988).

Plants are the building blocks of all life on the planet.

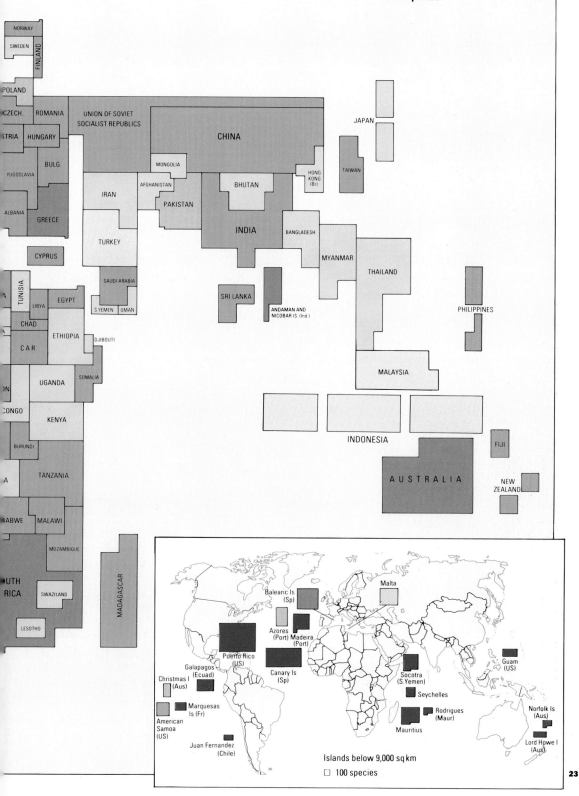

NORWAY

SWEDEN

FINLAND

POLAND

CZECH.

ROMANIA

UNION OF SOVIET SOCIALIST REPUBLICS

JAPAN

STRIA

HUNGARY

CHINA

BULG.

MONGOLIA

TAIWAN

YUGOSLAVIA

AFGHANISTAN

BHUTAN

HONG KONG (Br)

ALBANIA

IRAN

PAKISTAN

GREECE

TURKEY

INDIA

BANGLADESH

MYANMAR

CYPRUS

THAILAND

SAUDI ARABIA

TUNISIA

EGYPT

LIBYA

S.YEMEN

OMAN

SRI LANKA

ANDAMAN AND NICOBAR IS. (Ind.)

PHILIPPINES

CHAD

ETHIOPIA

DJIBOUTI

C A R

MALAYSIA

UGANDA

SOMALIA

ON

CONGO

KENYA

FIJI

BURUNDI

INDONESIA

A U S T R A L I A

NEW ZEALAND

TANZANIA

ABWE

MALAWI

MOZAMBIQUE

MADAGASCAR

UTH RICA

SWAZILAND

LESOTHO

Balearic Is (Sp)

Malta

Azores (Port)

Madeira (Port)

Puerto Rico (US)

Guam (US)

Galapagos (Ecuad)

Canary Is (Sp)

Socotra (S.Yemen)

Christmas I (Aus)

Seychelles

Marquesas Is (Fr)

Rodrigues (Maur)

Norfolk Is (Aus)

American Samoa (US)

Mauritius

Lord Howe I (Aus)

Juan Fernandez (Chile)

Islands below 9,000 sq km

☐ 100 species

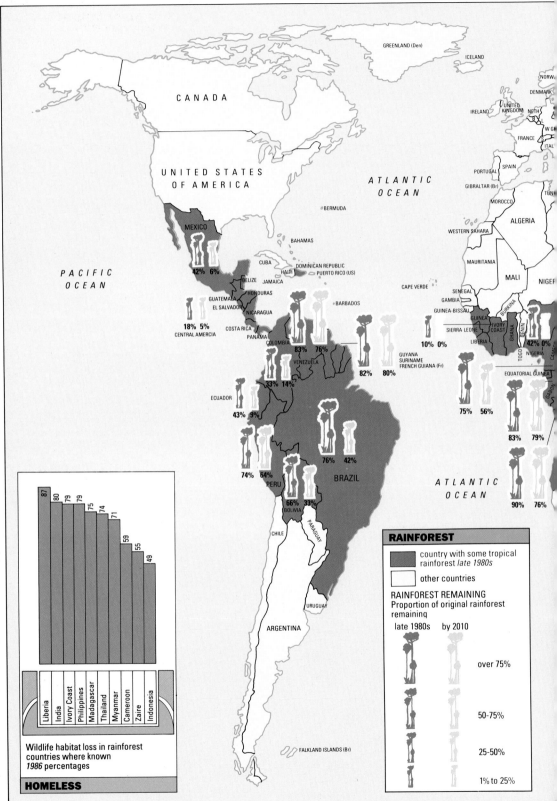

Map content (labels and data):

CANADA

GREENLAND (Den)
ICELAND

UNITED STATES OF AMERICA

ATLANTIC OCEAN

NORW
DENMARK
IRELAND · UNITED KINGDOM · NETH
W G
FRANCE
ITAL
PORTUGAL · SPAIN
GIBRALTAR (Br)
MOROCCO
WESTERN SAHARA
ALGERIA
TUN

PACIFIC OCEAN

MEXICO
42% 6%

BAHAMAS
CUBA
DOMINICAN REPUBLIC
HAITI · PUERTO RICO (US)
BELIZE · JAMAICA
GUATEMALA · HONDURAS
EL SALVADOR
NICARAGUA
COSTA RICA
PANAMA

BERMUDA

BARBADOS

CAPE VERDE

MAURITANIA
MALI
NIGE

SENEGAL
GAMBIA
GUINEA-BISSAU
GUINEA · BURKINA
SIERRA LEONE · IVORY COAST · GHANA · BENIN · TOGO
LIBERIA
NIGERIA · CAME

18% 5%
CENTRAL AMERICA

COLOMBIA
83% 76%
VENEZUELA

GUYANA
SURINAME
FRENCH GUIANA (Fr)
82% 80%

10% 0%

42% 0%

EQUATORIAL GUINEA
GABON

ECUADOR
33% 14%
43% 9%

75% 56%

83% 79%

74% 64%
PERU

76% 42%
BRAZIL

ATLANTIC OCEAN

90% 76%

66% 33%
BOLIVIA

33%
PARAGUAY

CHILE

URUGUAY

ARGENTINA

FALKLAND ISLANDS (Br)

RAINFOREST

country with some tropical rainforest *late 1980s*

other countries

RAINFOREST REMAINING
Proportion of original rainforest remaining

late 1980s   by 2010

over 75%

50-75%

25-50%

1% to 25%

Homeless bar chart:

Liberia 87
India 80
Ivory Coast 79
Philippines 79
Madagascar 75
Thailand 74
Myanmar 71
Cameroon 59
Zaire 55
Indonesia 49

Wildlife habitat loss in rainforest countries where known
1986 percentages

HOMELESS

Copyright © Swanston Publishing Limited

24

Data compiled by Friends of the Earth/Christine Lancaster   Main sources: Campbell & Hammond (1989); Myers & Houghton (1989).

A hundred acres of tropical forest are destroyed every minute.

UNION OF SOVIET SOCIALIST REPUBLICS

MONGOLIA

TURKEY

CYPRUS SYRIA
LEBANON
ISRAEL
JORDAN
IRAQ
IRAN
KUWAIT
BAHRAIN
QATAR
U.A.E
SAUDI ARABIA
OMAN
N. YEMEN
S. YEMEN
DJIBOUTI

AFGHANISTAN

PAKISTAN

CHINA

N. KOREA
S. KOREA
JAPAN

BHUTAN
NEPAL
B'DESH
MYANMAR
LAOS
THAILAND
CAM
VIETNAM

TAIWAN

PHILIPPINES

PACIFIC OCEAN

11% 6%
INDIA

49% 17%

23% 0%

20% 4%

SRI LANKA

17% 0%

BRUNEI

51% 20%
MALAYSIA

ETHIOPIA

SOMALIA

UGANDA
KENYA

MALDIVES

TANZANIA

SEYCHELLES

I N D O N E S I A

57% 38%

85% 68%

PAPUA NEW GUINEA

SOLOMON ISLANDS

INDIAN OCEAN

MALAWI
COMOROS
MADAGASCAR

32% 0%

MOZAMBIQUE

WESTERN SAMOA

FIJI

AUSTRALIA

NEW ZEALAND

THE GEOGRAPHY OF TROPICAL RAINFORESTS

original extent of rainforest

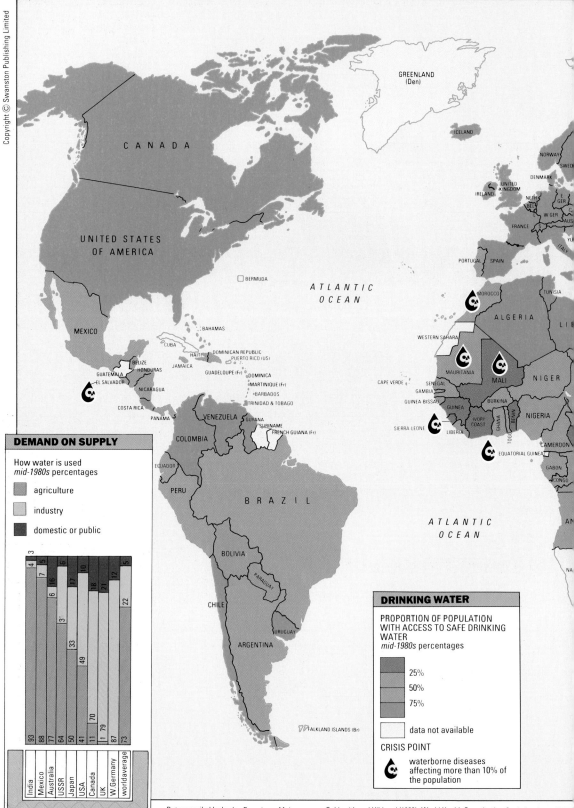

GREENLAND
(Den)

ICELAND

NORWAY

SWED

DENMARK

UNITED
KINGDOM

IRELAND

NETH

BEL

E
GER

W GER

AUS

FRANCE

ITAL

PORTUGAL   SPAIN

C A N A D A

UNITED STATES
OF AMERICA

☐ BERMUDA

A T L A N T I C
O C E A N

MEXICO

•BAHAMAS

CUBA

HAITI

DOMINICAN REPUBLIC
PUERTO RICO (US)

JAMAICA

GUADELOUPE (Fr)

•DOMINICA

•MARTINIQUE (Fr)

•BARBADOS

TRINIDAD & TOBAGO

BELIZE

GUATEMALA

HONDURAS

EL SALVADOR

NICARAGUA

COSTA RICA

PANAMA

VENEZUELA

GUYANA

SURINAME

FRENCH GUIANA (Fr)

COLOMBIA

ECUADOR

PERU

B R A Z I L

BOLIVIA

PARAGUAY

CHILE

URUGUAY

ARGENTINA

FALKLAND ISLANDS (Br)

MOROCCO

TUNISIA

ALGERIA

LI

WESTERN SAHARA

MAURITANIA

MALI

NIGER

CAPE VERDE

SENEGAL

GAMBIA

GUINEA-BISSAU

GUINEA

BURKINA

SIERRA LEONE

LIBERIA

IVORY
COAST

GHANA

BENIN

TOGO

NIGERIA

CAMEROON

EQUATORIAL GUINEA

GABON

CONGO

A T L A N T I C
O C E A N

AN

NA

DEMAND ON SUPPLY

How water is used
mid-1980s percentages

- agriculture
- industry
- domestic or public

	India	Mexico	Australia	USSR	Japan	USA	Canada	UK	W Germany	worldaverage
domestic	3	4	5	6		10	18	21	12	5
industry		7	16		33	49	70	1 79		22
agriculture	93	88	77	64	50	41	11	1	87	73

DRINKING WATER

PROPORTION OF POPULATION
WITH ACCESS TO SAFE DRINKING
WATER
mid-1980s percentages

- 25%
- 50%
- 75%
- data not available

CRISIS POINT

waterborne diseases
affecting more than 10% of
the population

Data compiled by Lesley Downing   Main sources: Goldsmith and Hildyard (1988); World Health Organization *Statistics Annual* (198

The availability of safe drinking water cannot be taken for granted, even in the 'developed' world.

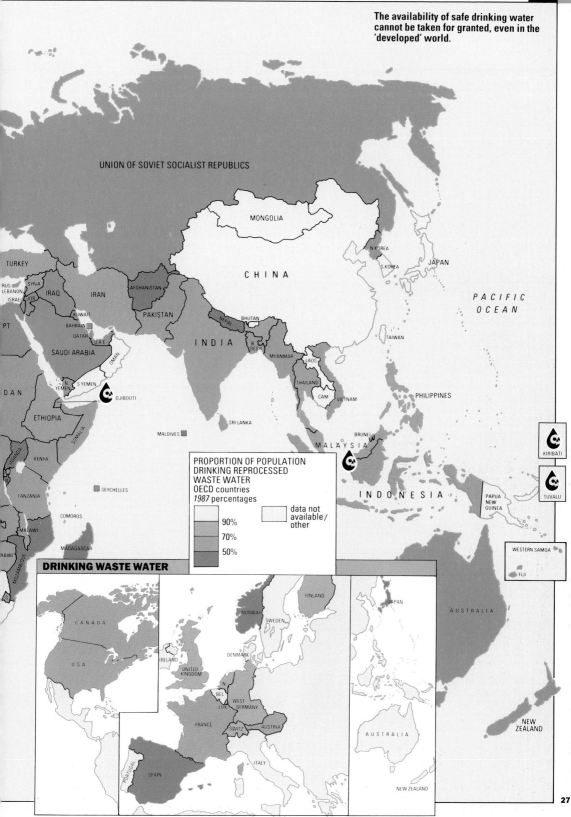

UNION OF SOVIET SOCIALIST REPUBLICS

MONGOLIA

CHINA

N KOREA

S KOREA

JAPAN

TURKEY

RUS
LEBANON
ISRAEL
JOR
IRAQ

SYRIA

IRAN

AFGHANISTAN

PT

KUWAIT

BAHRAIN

QATAR

UAE

PAKISTAN

NEPAL

BHUTAN

SAUDI ARABIA

OMAN

INDIA

B
DESH

MYANMAR

LAOS

TAIWAN

PACIFIC
OCEAN

N
YEMEN

S YEMEN

DJIBOUTI

THAILAND

CAM

VIETNAM

PHILIPPINES

D A N

ETHIOPIA

SOMALIA

SRI LANKA

MALDIVES

BRUNEI

MALAYSIA

KIRIBATI

UGANDA

KENYA

SEYCHELLES

I N D O N E S I A

TUVALU

TANZANIA

COMOROS

PAPUA
NEW
GUINEA

MALAWI

MADAGASCAR

WESTERN SAMOA

ABWE

MOZAMBIQUE

FIJI

PROPORTION OF POPULATION DRINKING REPROCESSED WASTE WATER
OECD countries
1987 percentages

	data not available/ other
90%	
70%	
50%	

DRINKING WASTE WATER

AUSTRALIA

CANADA

FINLAND

JAPAN

NORWAY

SWEDEN

U S A

IRELAND

DENMARK

UNITED
KINGDOM

BEL

LUX

N

WEST
GERMANY

FRANCE

SWITZ

AUSTRIA

AUSTRALIA

PORTUGAL

SPAIN

ITALY

NEW ZEALAND

NEW
ZEALAND

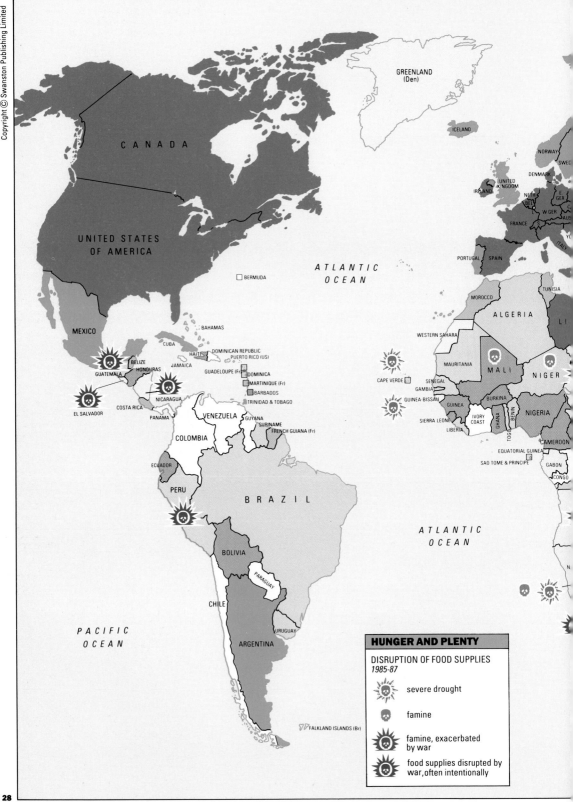

GREENLAND
(Den)

ICELAND

NORWAY

SWED

DENMARK

UNITED
KINGDOM

IRELAND

NETH.
BEL.

GER

W GER

AUS

FRANCE

ITALY

PORTUGAL   SPAIN

TUNISIA

MOROCCO

ALGERIA

LI

WESTERN SAHARA

MAURITANIA

MALI

NIGER

CAPE VERDE

SENEGAL

GAMBIA

GUINEA-BISSAU

BURKINA

SIERRA LEONE

GUINEA

IVORY
COAST

GHANA

BENIN

TOGO

NIGERIA

LIBERIA

CAMEROON

EQUATORIAL GUINEA

SAO TOME & PRINCIPE

GABON

CONGO

CANADA

UNITED STATES
OF AMERICA

ATLANTIC
OCEAN

BERMUDA

BAHAMAS

MEXICO

CUBA

HAITI

DOMINICAN REPUBLIC
PUERTO RICO (US)

JAMAICA

BELIZE

GUATEMALA

HONDURAS

GUADELOUPE (Fr)

DOMINICA

MARTINIQUE (Fr)

BARBADOS

TRINIDAD & TOBAGO

EL SALVADOR

NICARAGUA

COSTA RICA

PANAMA

VENEZUELA

GUYANA

SURINAME

FRENCH GUIANA (Fr)

COLOMBIA

ECUADOR

PERU

BRAZIL

ATLANTIC
OCEAN

BOLIVIA

PARAGUAY

CHILE

PACIFIC
OCEAN

ARGENTINA

URUGUAY

FALKLAND ISLANDS (Br)

HUNGER AND PLENTY

DISRUPTION OF FOOD SUPPLIES
1985-87

severe drought

famine

famine, exacerbated
by war

food supplies disrupted by
war, often intentionally

28

Data compiled by Jeanne Kasperson and Joni Seager   Main sources: FAO *Production Statistics* (1987); Kates (1989).

The persistence of hunger in a world of plenty represents the failure of social and economic systems.

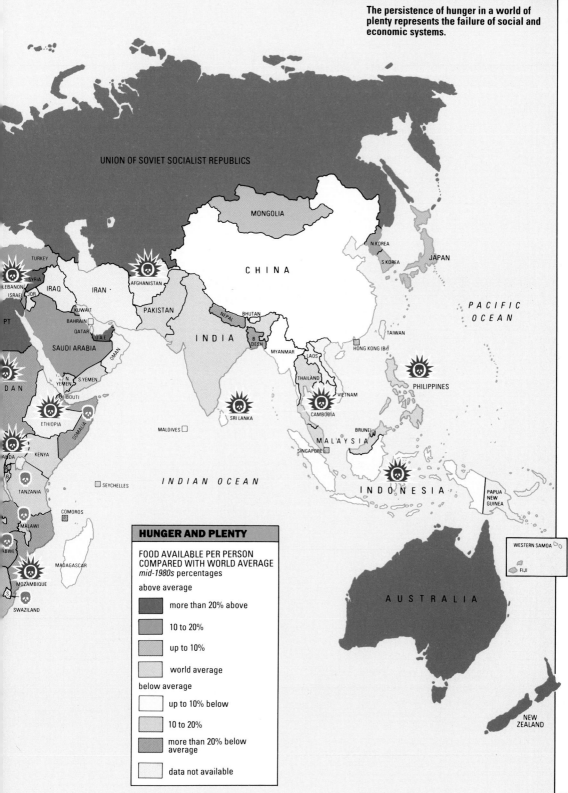

UNION OF SOVIET SOCIALIST REPUBLICS

MONGOLIA

CHINA

N KOREA

S KOREA

JAPAN

PACIFIC OCEAN

TURKEY

LEBANON
SYRIA
ISRAEL JOR
IRAQ

IRAN

AFGHANISTAN

KUWAIT
BAHRAIN
QATAR
UAE

PT

SAUDI ARABIA

OMAN

N
YEMEN S YEMEN

PAKISTAN

NEPAL

BHUTAN

INDIA

B DESH

MYANMAR

LAOS

THAILAND

TAIWAN

HONG KONG (Br)

VIETNAM

CAMBODIA

PHILIPPINES

DAN

DJIBOUTI

ETHIOPIA

SOMALIA

MALDIVES

SRI LANKA

BRUNEI

MALAYSIA

SINGAPORE

ANDA

KENYA

INDIAN OCEAN

SEYCHELLES

INDONESIA

PAPUA
NEW
GUINEA

B

TANZANIA

COMOROS

MALAWI

ABWE

MADAGASCAR

WESTERN SAMOA

FIJI

AUSTRALIA

MOZAMBIQUE

S

SWAZILAND

NEW
ZEALAND

HUNGER AND PLENTY

FOOD AVAILABLE PER PERSON
COMPARED WITH WORLD AVERAGE
mid-1980s percentages

above average

more than 20% above

10 to 20%

up to 10%

world average

below average

up to 10% below

10 to 20%

more than 20% below
average

data not available

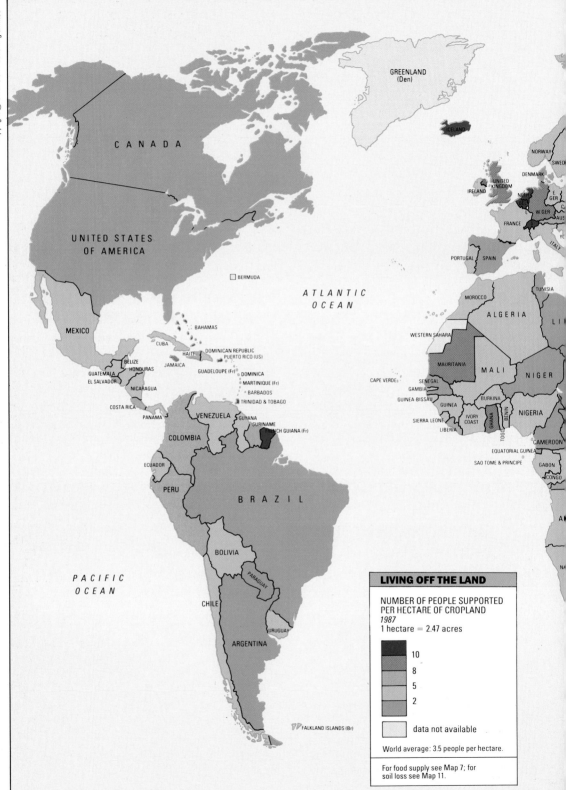

CANADA

UNITED STATES
OF AMERICA

MEXICO

BERMUDA

ATLANTIC
OCEAN

BAHAMAS

CUBA

BELIZE
GUATEMALA   HONDURAS   JAMAICA
EL SALVADOR   HAITI   DOMINICAN REPUBLIC
NICARAGUA   GUADELOUPE (Fr)   PUERTO RICO (US)
                    DOMINICA
COSTA RICA   MARTINIQUE (Fr)
PANAMA   BARBADOS
        VENEZUELA   TRINIDAD & TOBAGO
COLOMBIA   GUYANA
        SURINAME
ECUADOR   FRENCH GUIANA (Fr)

PERU

B R A Z I L

BOLIVIA

PARAGUAY

PACIFIC
OCEAN

CHILE

URUGUAY

ARGENTINA

FALKLAND ISLANDS (Br)

GREENLAND
(Den)

ICELAND

NORWAY   SWEDEN
        DENMARK
IRELAND   UNITED
        KINGDOM
        NETH
        BEL   E
        W GER   GER
            C
        AUS
FRANCE   YU
                ITALY

PORTUGAL   SPAIN

                TUNISIA
MOROCCO
        ALGERIA   LI
WESTERN SAHARA

        MAURITANIA   MALI   NIGER
CAPE VERDE
        SENEGAL
GAMBIA   BURKINA   BENIN
GUINEA-BISSAU   GUINEA   NIGERIA
                IVORY   GHANA
SIERRA LEONE   COAST   TOGO
        LIBERIA   CAMEROON
EQUATORIAL GUINEA
SAO TOME & PRINCIPE   GABON
                CONGO
                    A
                    NA

LIVING OFF THE LAND

NUMBER OF PEOPLE SUPPORTED
PER HECTARE OF CROPLAND
1987
1 hectare = 2.47 acres

- 10
- 8
- 5
- 2

data not available

World average: 3.5 people per hectare.

For food supply see Map 7; for
soil loss see Map 11.

Data compiled by Michael Shanabrook   Main sources: FAO, *Production Statistics Yearbook* (1989); Gramfors (1989); Kurian (1984).

Agricultural land per person is shrinking the world over, although pressure on the land is not evenly distributed.

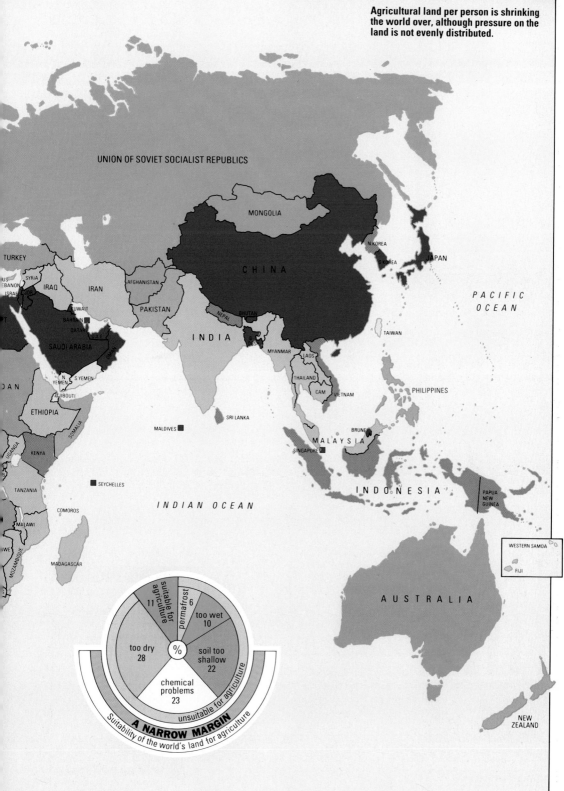

UNION OF SOVIET SOCIALIST REPUBLICS

MONGOLIA

N.KOREA

S.KOREA

JAPAN

CHINA

TURKEY

SYRIA

LEBANON

ISRAEL

IRAQ

IRAN

AFGHANISTAN

KUWAIT

BAHRAIN

QATAR

U.A.E.

PAKISTAN

NEPAL

BHUTAN

OMAN

SAUDI ARABIA

INDIA

MYANMAR

LAOS

N.YEMEN

S.YEMEN

THAILAND

DJIBOUTI

CAM

VIETNAM

PHILIPPINES

ETHIOPIA

SOMALIA

TAIWAN

PACIFIC OCEAN

SRI LANKA

MALDIVES

BRUNEI

MALAYSIA

UGANDA

KENYA

SINGAPORE

TANZANIA

SEYCHELLES

INDONESIA

PAPUA NEW GUINEA

COMOROS

INDIAN OCEAN

MALAWI

MOZAMBIQUE

MADAGASCAR

AUSTRALIA

WESTERN SAMOA

FIJI

NEW ZEALAND

A NARROW MARGIN
Suitability of the world's land for agriculture

suitable for agriculture 11

permafrost 6

too wet 10

too dry 28

soil too shallow 22

chemical problems 23

%

unsuitable for agriculture

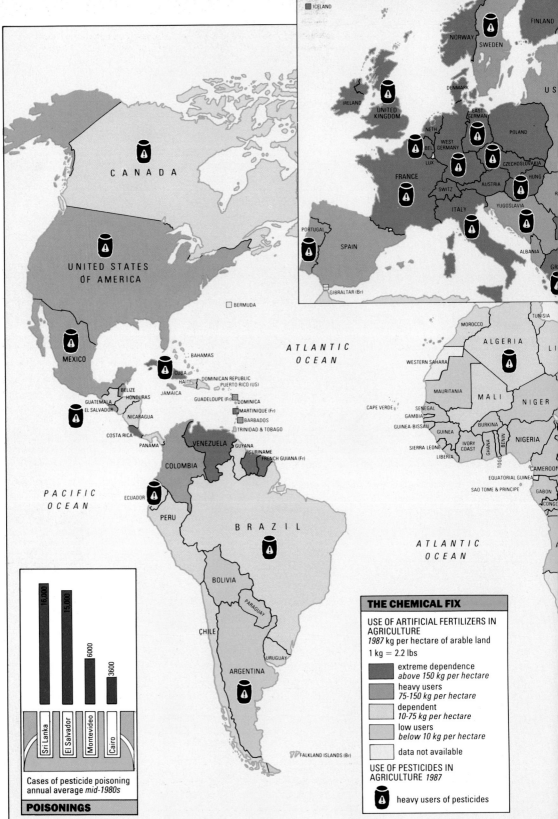

CANADA

UNITED STATES
OF AMERICA

MEXICO

GUATEMALA
EL SALVADOR

COSTA RICA
PANAMA

ECUADOR

PERU

BRAZIL

BOLIVIA

PARAGUAY

CHILE

URUGUAY

ARGENTINA

BERMUDA

BAHAMAS

CUBA
HAITI
DOMINICAN REPUBLIC
PUERTO RICO (US)
JAMAICA
GUADELOUPE (Fr)
DOMINICA
MARTINIQUE (Fr)
BARBADOS
TRINIDAD & TOBAGO

BELIZE
HONDURAS
NICARAGUA

VENEZUELA GUYANA
SURINAME
FRENCH GUIANA (Fr)
COLOMBIA

FALKLAND ISLANDS (Br)

PACIFIC
OCEAN

ATLANTIC
OCEAN

ATLANTIC
OCEAN

ICELAND

NORWAY
SWEDEN

FINLAND

IRELAND
UNITED
KINGDOM

DENMARK

US

NETH
BEL
LUX
WEST
GERMANY

EAST
GERMANY

POLAND

CZECHOSLOVAKIA

FRANCE

SWITZ
AUSTRIA

HUNG

ITALY

YUGOSLAVIA

R

ALBANIA

GR

PORTUGAL

SPAIN

GIBRALTAR (Br)

MOROCCO

TUNISIA

WESTERN SAHARA

ALGERIA

LI

MAURITANIA

MALI

NIGER

CAPE VERDE

SENEGAL
GAMBIA
GUINEA-BISSAU
GUINEA
SIERRA LEONE
LIBERIA

BURKINA

IVORY
COAST
GHANA
TOGO
BENIN

NIGERIA

CAMEROON

EQUATORIAL GUINEA
SAO TOME & PRINCIPE

GABON
CONGO

Cases of pesticide poisoning
annual average *mid-1980s*

16,000 — Sri Lanka
15,000 — El Salvador
6000 — Montevideo
3600 — Cairo

POISONINGS

THE CHEMICAL FIX

USE OF ARTIFICIAL FERTILIZERS IN
AGRICULTURE
1987 kg per hectare of arable land
1 kg = 2.2 lbs

- extreme dependence
  above 150 kg per hectare
- heavy users
  75-150 kg per hectare
- dependent
  10-75 kg per hectare
- low users
  below 10 kg per hectare
- data not available

USE OF PESTICIDES IN
AGRICULTURE *1987*

heavy users of pesticides

Data compiled by Joni Seager   Main sources: FAO *Fertilizer Yearbook* (1988); FAO *Production Yearbook* (1987); FAO *Trade Statistics* (1988).

The more agricultural chemicals are used, the more they are needed.

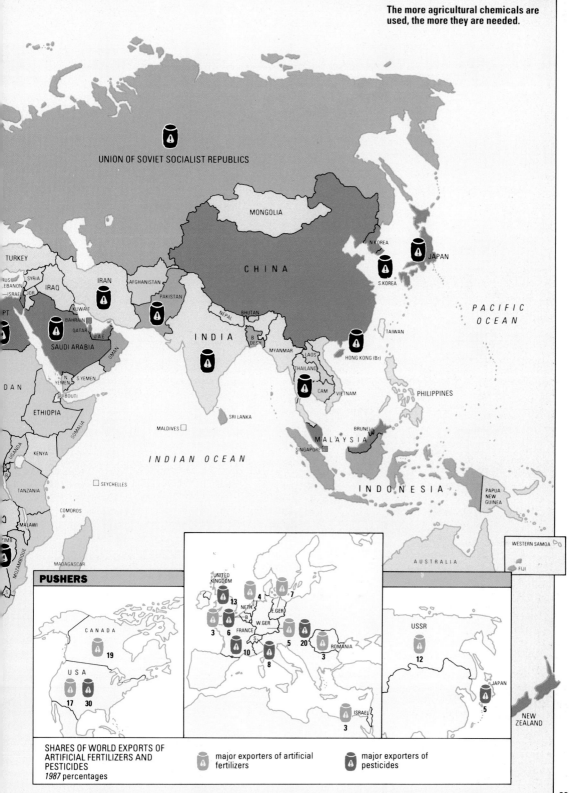

UNION OF SOVIET SOCIALIST REPUBLICS

TURKEY

RUS
SYRIA
LEBANON
ISRAEL
JOR
IRAQ

IRAN

AFGHANISTAN

PT

KUWAIT
BAHRAIN
QATAR
U.A.E.
SAUDI ARABIA
OMAN

PAKISTAN

NEPAL

BHUTAN

INDIA

B
DESH

MYANMAR

LAOS

THAILAND

CAM

VIETNAM

MONGOLIA

CHINA

N.KOREA

JAPAN

S.KOREA

TAIWAN

HONG KONG (Br)

PACIFIC
OCEAN

PHILIPPINES

DAN

I N
YEMEN
S.YEMEN

DJIBOUTI

ETHIOPIA

SOMALIA

SRI LANKA

MALDIVES

INDIAN OCEAN

BRUNEI

MALAYSIA

SINGAPORE

INDONESIA

UGANDA

KENYA

TANZANIA

COMOROS

SEYCHELLES

PAPUA
NEW
GUINEA

MALAWI

MOZAMBIQUE

IMB

MADAGASCAR

AUSTRALIA

WESTERN SAMOA

FIJI

PUSHERS

CANADA   19

USA
17   30

UNITED
KINGDOM
13   4   7

NETH
B
W.GER   3
FRANCE   5   20
3   6
10
8   5
ROMANIA   3

E.GER

ISRAEL
3

USSR

12

JAPAN

5

NEW
ZEALAND

SHARES OF WORLD EXPORTS OF
ARTIFICIAL FERTILIZERS AND
PESTICIDES
1987 percentages

major exporters of artificial
fertilizers

major exporters of
pesticides

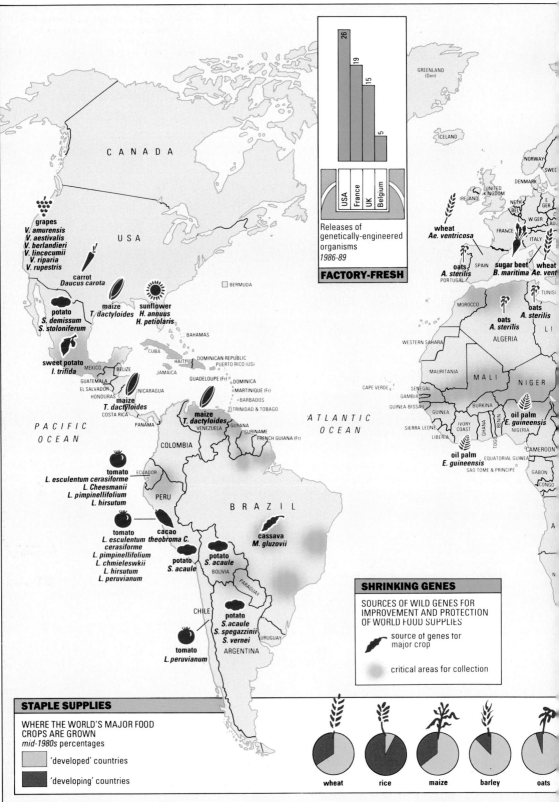

CANADA

USA

grapes
V. amurensis
V. aestivalis
V. berlandieri
V. lincecumii
V. riparia
V. rupestris

carrot
Daucus carota

maize
T. dactyloides

sunflower
H. annuus
H. petiolaris

potato
S. demissum
S. stoloniferum

sweet potato
I. trifida

BERMUDA

BAHAMAS

CUBA
JAMAICA
HAITI
DOMINICAN REPUBLIC
PUERTO RICO (US)

MEXICO BELIZE
GUATEMALA
EL SALVADOR
HONDURAS
NICARAGUA
COSTA RICA PANAMA

maize
T. dactyloides

GUADELOUPE (Fr)
DOMINICA
MARTINIQUE (Fr)
BARBADOS
TRINIDAD & TOBAGO

maize
T. dactyloides
VENEZUELA GUYANA
SURINAME
FRENCH GUIANA (Fr)

COLOMBIA

ECUADOR

tomato
L. esculentum cerasiforme
L. Cheesmanii
L. pimpinellifolium
L. hirsutum

PERU

BRAZIL

tomato
L. esculentum
cerasiforme
L. pimpinellifolium
L. chmieleswkii
L. hirsutum
L. peruvianum

cacao
theobroma C.

cassava
M. gluzovii

potato
S. acaule

potato
S. acaule

BOLIVIA

PARAGUAY

CHILE

potato
S. acaule
S. spegazzinii
S. vernei

URUGUAY

ARGENTINA

tomato
L. peruvianum

PACIFIC
OCEAN

ATLANTIC
OCEAN

GREENLAND
(Den)

ICELAND

NORWAY
SWE
DENMARK
IRELAND UNITED
KINGDOM
NETH
BEL
W GER
E
GER
AU
FRANCE
ITALY

SPAIN
PORTUGAL

wheat
Ae. ventricosa

oats
A. sterilis

sugar beet
B. maritima

wheat
Ae. vent

TUNISI

MOROCCO

oats
A. sterilis

oats
A. sterilis

WESTERN SAHARA

ALGERIA

LI

MAURITANIA

MALI

NIGER

CAPE VERDE
SENEGAL
GAMBIA
GUINEA-BISSAU
GUINEA
SIERRA LEONE
LIBERIA

BURKINA

IVORY
COAST
GHANA
TOGO
BENIN
NIGERIA

oil palm
E. guineensis

CAMEROON

oil palm
E. guineensis

EQUATORIAL GUINEA
SAO TOME & PRINCIPE

GABON
CONGO

FACTORY-FRESH

26	19	15	5
USA	France	UK	Belgium

Releases of
genetically-engineered
organisms
1986-89

SHRINKING GENES

SOURCES OF WILD GENES FOR
IMPROVEMENT AND PROTECTION
OF WORLD FOOD SUPPLIES

source of genes for
major crop

critical areas for collection

STAPLE SUPPLIES

WHERE THE WORLD'S MAJOR FOOD
CROPS ARE GROWN
mid-1980s percentages

'developed' countries

'developing' countries

wheat   rice   maize   barley   oats

Data compiled by Simon Albrecht   Main sources: Prescott-Allen (1988); Myers (1984)

There is no such thing as self-sufficiency in genetic resources. The world's food supplies depend on cooperative protection of the global gene pool of wild plants.

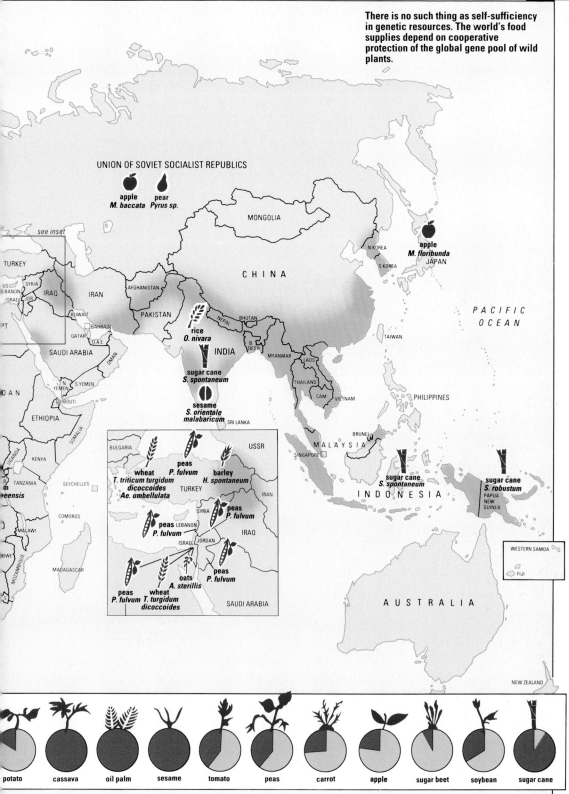

UNION OF SOVIET SOCIALIST REPUBLICS

apple
M. baccata

pear
Pyrus sp.

MONGOLIA

see inset

TURKEY

US
LEBANON
SYRIA
ISRAEL JOR.
IRAQ
IRAN

AFGHANISTAN

CHINA

N. KOREA
S. KOREA

apple
M. floribunda
JAPAN

PT

KUWAIT
BAHRAIN
QATAR
U.A.E.

PAKISTAN

NEPAL
BHUTAN

rice
O. nivara

INDIA

B.
DESH

MYANMAR

LAOS

TAIWAN

*PACIFIC
OCEAN*

SAUDI ARABIA

OMAN

sugar cane
S. spontaneum

THAILAND

CAM
VIETNAM

PHILIPPINES

N.
YEMEN
S. YEMEN

DJIBOUTI

sesame
*S. orientale
malabaricum* SRI LANKA

DAN

ETHIOPIA

SOMALIA

UGANDA

KENYA

BULGARIA

USSR

peas
P. fulvum
barley
H. spontaneum

BRUNEI

MALAYSIA

SINGAPORE

TANZANIA

SEYCHELLES

B.
eensis

wheat
*T. triticum turgidum
dicoccoides
Ae. umbellulata*

TURKEY

IRAN

sugar cane
S. spontaneum

INDONESIA

sugar cane
S. robustum
PAPUA
NEW
GUINEA

COMOROS

SYRIA
peas
P. fulvum

MALAWI

peas LEBANON
P. fulvum

ISRAEL JORDAN

IRAQ

WESTERN SAMOA

ZWE

MADAGASCAR

peas
P. fulvum wheat
oats
A. sterilis

peas
P. fulvum

FIJI

MOZAMBIQUE

S

peas
P. fulvum wheat *T. turgidum
dicoccoides*

SAUDI ARABIA

AUSTRALIA

NEW ZEALAND

potato cassava oil palm sesame tomato peas carrot apple sugar beet soybean sugar cane

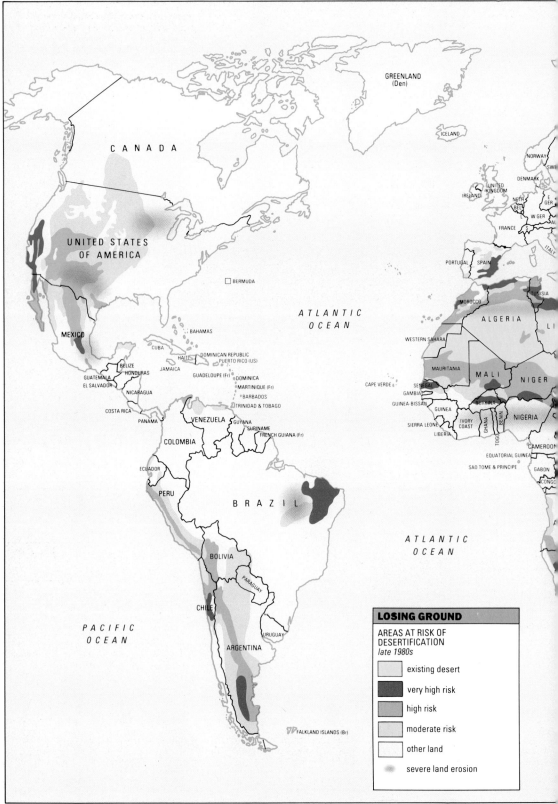

LOSING GROUND

AREAS AT RISK OF
DESERTIFICATION
late 1980s

existing desert

very high risk

high risk

moderate risk

other land

severe land erosion

Data compiled by Andy Crump   Main sources: UNEP *World Conservation Strategy* (1980); World Resources Institute (1988); Myers (1984).

By the year 2000 one-third of the world's agricultural land will have turned to dust.

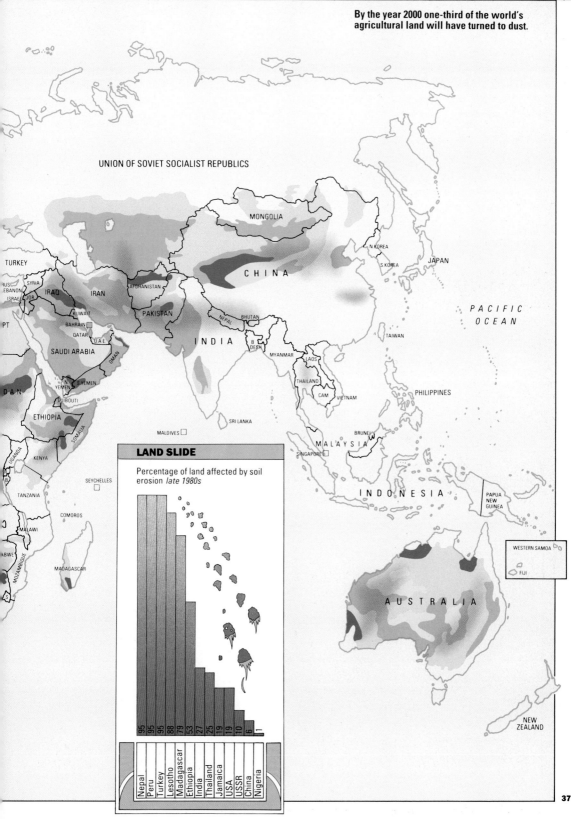

UNION OF SOVIET SOCIALIST REPUBLICS

MONGOLIA

N KOREA

S KOREA

JAPAN

TURKEY

RUS.
LEBANON
ISRAEL
SYRIA
JOR
IRAQ
IRAN
AFGHANISTAN

PT

KUWAIT
BAHRAIN
QATAR
U.A.E
OMAN

SAUDI ARABIA

PAKISTAN

NEPAL

BHUTAN

CHINA

TAIWAN

PACIFIC
OCEAN

INDIA

B
DESH

MYANMAR

LAOS

THAILAND

CAM

VIETNAM

PHILIPPINES

SRI LANKA

DAN

YEMEN
S.YEMEN

DJIBOUTI

ETHIOPIA

SOMALIA

MALDIVES

BRUNEI

MALAYSIA

SINGAPORE

UGANDA

KENYA

SEYCHELLES

INDONESIA

PAPUA
NEW
GUINEA

TANZANIA

COMOROS

MALAWI

ABWE

MOZAMBIQUE

MADAGASCAR

AUSTRALIA

WESTERN SAMOA

FIJI

NEW
ZEALAND

LAND SLIDE

Percentage of land affected by soil erosion *late 1980s*

95	95	95	88	79	53	27	25	19	19	10	6	1
Nepal	Peru	Turkey	Lesotho	Madagascar	Ethiopia	India	Thailand	Jamaica	USA	USSR	China	Nigeria

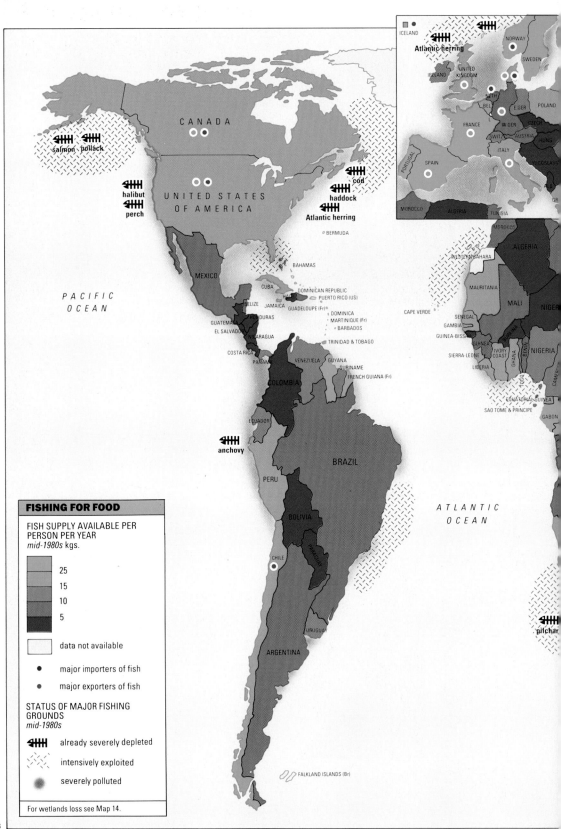

ICELAND

Atlantic herring

NORWAY

SWEDEN

IRELAND

UNITED
KINGDOM

NETH.
BEL.
E GER
W GER

POLAND

FRANCE

SWITZ.
ITALY

AUSTRIA
HUNG.

CZECH.

YUGOSLAVIA

PORTUGAL

SPAIN

GR.

MOROCCO

ALGERIA

TUNISIA

CANADA

salmon   pollack

halibut

perch

UNITED STATES
OF AMERICA

cod

haddock

Atlantic herring

MOROCCO

ALGERIA

WESTERN SAHARA

BERMUDA

MEXICO

PACIFIC
OCEAN

BAHAMAS

CUBA

DOMINICAN REPUBLIC

PUERTO RICO (US)

HAITI

JAMAICA

GUADELOUPE (Fr)

DOMINICA

MARTINIQUE (Fr)

BARBADOS

TRINIDAD & TOBAGO

BELIZE

GUATEMALA

HONDURAS

EL SALVADOR

NICARAGUA

COSTA RICA

PANAMA

VENEZUELA

GUYANA

SURINAME

FRENCH GUIANA (Fr)

COLOMBIA

ECUADOR

anchovy

BRAZIL

PERU

BOLIVIA

PARAGUAY

CHILE

URUGUAY

ARGENTINA

FALKLAND ISLANDS (Br)

CAPE VERDE

MAURITANIA

SENEGAL

GAMBIA

GUINEA-BISSAU

GUINEA

SIERRA LEONE

LIBERIA

IVORY
COAST

MALI

BURKINA

GHANA

TOGO

BENIN

NIGER

NIGERIA

CAMEROON

EQUATORIAL GUINEA

SAO TOME & PRINCIPE

GABON

ATLANTIC
OCEAN

pilchar

FISHING FOR FOOD

FISH SUPPLY AVAILABLE PER
PERSON PER YEAR
mid-1980s kgs.

25

15

10

5

data not available

● major importers of fish

● major exporters of fish

STATUS OF MAJOR FISHING
GROUNDS
mid-1980s

already severely depleted

intensively exploited

severely polluted

For wetlands loss see Map 14.

38

Data compiled by Andy Crump   Main sources: FAO *Fishery Statistics* (1986, 1987); Myers (1984).

Fisheries provide almost a quarter of the world's supply of animal protein, but face threats from many sources.

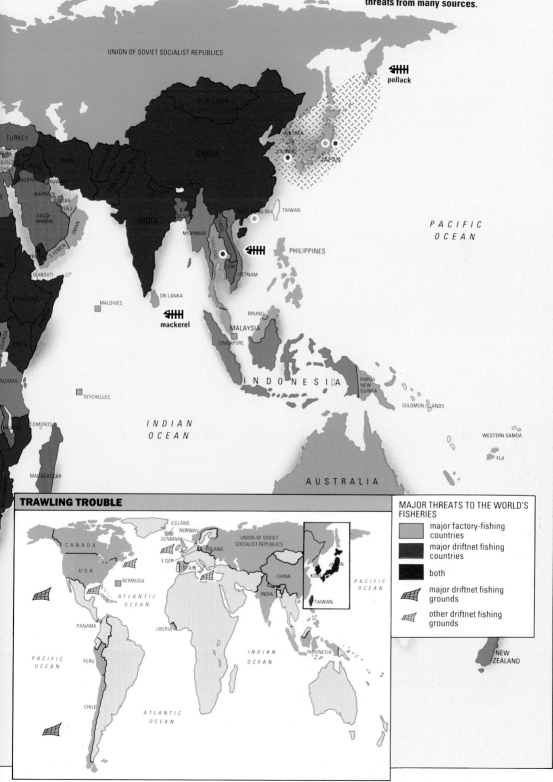

UNION OF SOVIET SOCIALIST REPUBLICS

pollack

MONGOLIA

TURKEY

RUS...
BAN...
ISRAEL
JORDAN
KUWAIT
BAHRAIN
QATAR
U.A.E
SAUDI ARABIA
OMAN
S.YEMEN
YEMEN
DJIBOUTI
ETHIOPIA
UGANDA
KENYA
NZANIA
COMOROS
SEYCHELLES
MADAGASCAR

IRAQ
IRAN

CHINA

N.KOREA
S.KOREA
JAPAN

INDIA
B...DESH
MYANMAR
LAOS
THAILAND
CAM...
VIETNAM
SRI LANKA
MALDIVES

HONG KONG (Br)
TAIWAN

PHILIPPINES

mackerel

BRUNEI
MALAYSIA
SINGAPORE

INDONESIA

PAPUA NEW GUINEA

SOLOMON ISLANDS

WESTERN SAMOA
FIJI

PACIFIC OCEAN

INDIAN OCEAN

AUSTRALIA

TRAWLING TROUBLE

CANADA

ICELAND
NORWAY
DENMARK

UNION OF SOVIET SOCIALIST REPUBLICS

POLAND
E GER
SPAIN

USA

BERMUDA

ATLANTIC OCEAN

PANAMA

LIBERIA

PERU

PACIFIC OCEAN

CHILE

ATLANTIC OCEAN

INDIAN OCEAN

CHINA
INDIA
KOREA
JAPAN
TAIWAN

PACIFIC OCEAN

INDONESIA

NEW ZEALAND

MAJOR THREATS TO THE WORLD'S FISHERIES

- major factory-fishing countries
- major driftnet fishing countries
- both
- major driftnet fishing grounds
- other driftnet fishing grounds

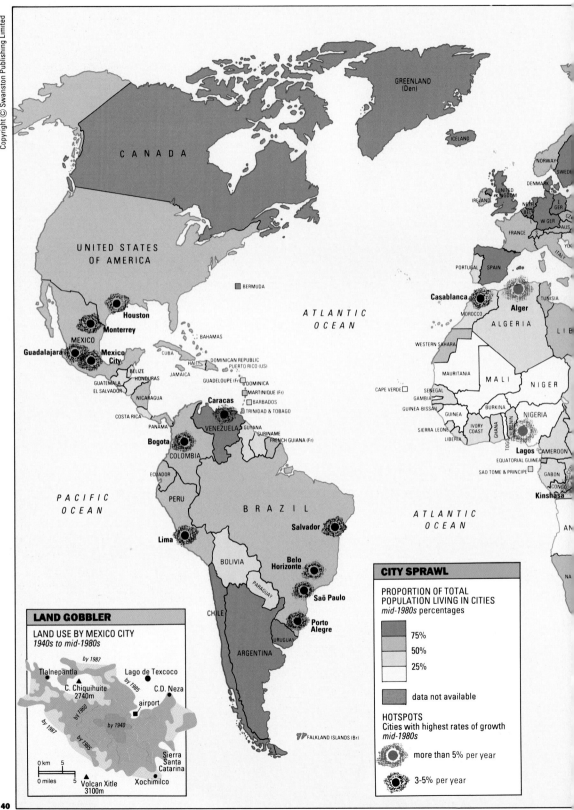

GREENLAND
(Den)

ICELAND

C A N A D A

NORWAY
SWEDE
DENMARK
UNITED
KINGDOM
IRELAND
NETH
GER
BEL
W GER
CZE
AUS
FRANCE
YU
ITALY

UNITED STATES
OF AMERICA

PORTUGAL SPAIN

BERMUDA

ATLANTIC
OCEAN

Casablanca
Alger
MOROCCO
TUNISIA
ALGERIA
LIB

Houston

WESTERN SAHARA

Monterrey

BAHAMAS

MEXICO
Guadalajara
MEXICO
City
CUBA
BELIZE
GUATEMALA
EL SALVADOR
HONDURAS
NICARAGUA
COSTA RICA
PANAMA

HAITI
DOMINICAN REPUBLIC
PUERTO RICO (US)
JAMAICA
GUADELOUPE (Fr)
DOMINICA
MARTINIQUE (Fr)
BARBADOS
TRINIDAD & TOBAGO

CAPE VERDE

MAURITANIA

MALI

NIGER

SENEGAL
GAMBIA
GUINEA BISSAU
GUINEA
SIERRA LEONE
LIBERIA
IVORY
COAST
BURKINA
GHANA
TOGO
BENIN
NIGERIA

Caracas
VENEZUELA
GUYANA
SURINAME
FRENCH GUIANA (Fr)

Bogota
COLOMBIA

Lagos
CAMEROON
EQUATORIAL GUINEA
SAO TOME & PRINCIPE
GABON
CONGO

ECUADOR

PACIFIC
OCEAN

PERU

B R A Z I L

ATLANTIC
OCEAN

Kinshasa

AN

Lima

Salvador

BOLIVIA

Belo
Horizonte

NA

CHILE

PARAGUAY

Saõ Paulo

Porto
Alegre

URUGUAY

ARGENTINA

FALKLAND ISLANDS (Br)

Data compiled by Michael Steinitz   Main sources: UN *Prospects of World Urbanization* (1988); *National Geographic* (1984).

CITY SPRAWL

PROPORTION OF TOTAL
POPULATION LIVING IN CITIES
mid-1980s percentages

75%

50%

25%

data not available

HOTSPOTS
Cities with highest rates of growth
mid-1980s

more than 5% per year

3-5% per year

LAND GOBBLER

LAND USE BY MEXICO CITY
1940s to mid-1980s

by 1987

Tlalnepantla

Lago de Texcoco

C. Chiquihuite
2740m

by 1985

C.D. Neza

airport

by 1960

by 1987

by 1940

by 1985

Sierra
Santa
Catarina

0 km   5

0 miles   5

Volcan Xitle
3100m

Xochimilco

40

By the year 2000, half the world's population will live in cities.

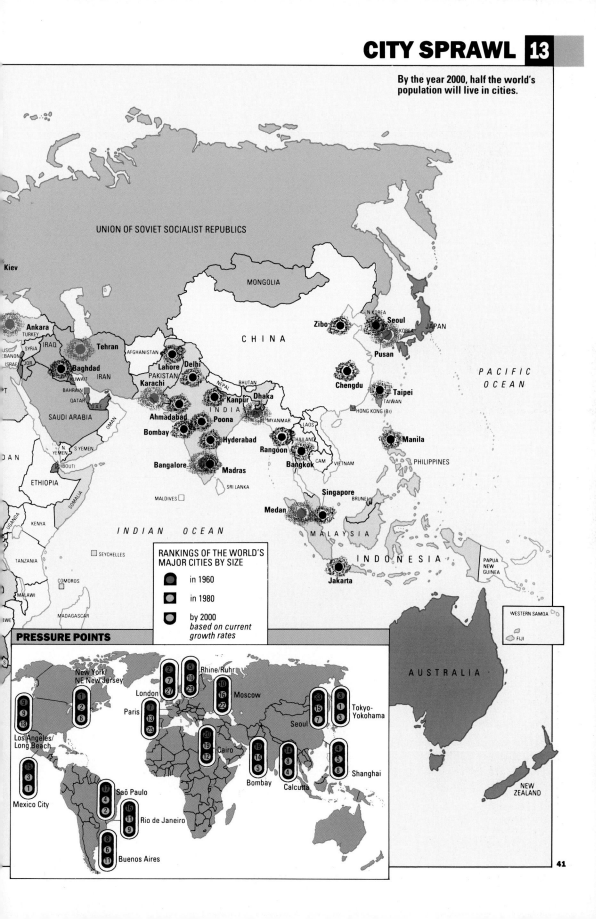

UNION OF SOVIET SOCIALIST REPUBLICS

Kiev

MONGOLIA

CHINA

Ankara
TURKEY
IRAQ
Tehran
Baghdad
IRAN
LEBANON
ISRAEL JOR
SAUDI ARABIA
BAHRAIN
QATAR
U.A.E.
OMAN
N. YEMEN
S YEMEN
DJIBOUTI

AFGHANISTAN
PAKISTAN
Lahore Delhi
Karachi
Kanpur Dhaka
NEPAL BHUTAN
Ahmadabad Poona
I N D I A
Bombay
Hyderabad
Bangalore Madras
SRI LANKA
MALDIVES

Zibo
N KOREA Seoul
S KOREA
Pusan
JAPAN

Chengdu
Taipei
TAIWAN
HONG KONG (Br)

MYANMAR
LAOS
Rangoon
THAILAND
Bangkok
CAM
VIETNAM

Manila
PHILIPPINES

PACIFIC
OCEAN

ETHIOPIA
UGANDA
KENYA
SOMALIA
TANZANIA

SEYCHELLES

I N D I A N   O C E A N

Singapore
SINGAPORE
Medan
M A L A Y S I A
BRUNEI

I N D O N E S I A

PAPUA
NEW
GUINEA

COMOROS
MALAWI
MADAGASCAR

Jakarta

RANKINGS OF THE WORLD'S MAJOR CITIES BY SIZE

- in 1960
- in 1980
- by 2000
  based on current growth rates

PRESSURE POINTS

New York/
NE New Jersey
Rhine/Ruhr
Moscow
London
Paris
Tokyo-
Yokohama
Seoul
Los Angeles/
Long Beach
Cairo
Shanghai
Mexico City
Bombay
Calcutta
São Paulo
Rio de Janeiro
Buenos Aires

WESTERN SAMOA

FIJI

A U S T R A L I A

NEW
ZEALAND

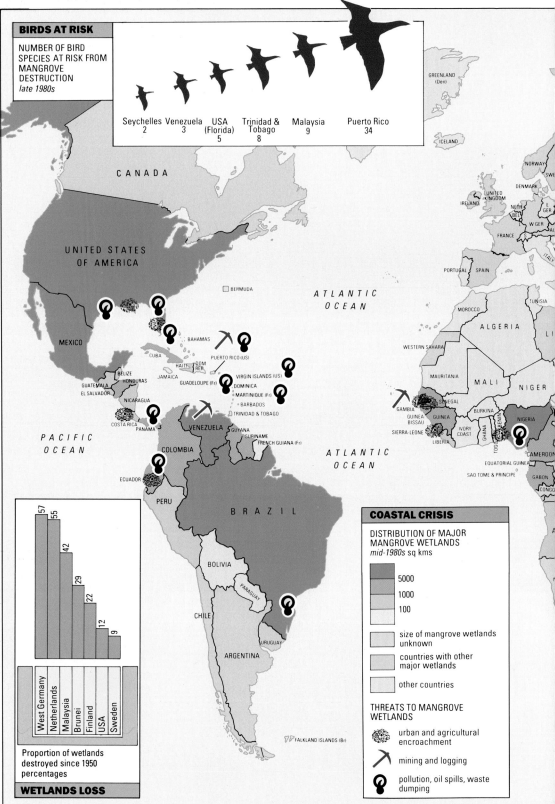

BIRDS AT RISK

NUMBER OF BIRD
SPECIES AT RISK FROM
MANGROVE
DESTRUCTION
late 1980s

Seychelles	Venezuela	USA (Florida)	Trinidad & Tobago	Malaysia	Puerto Rico
2	3	5	8	9	34

CANADA

GREENLAND (Den)

ICELAND

NORWAY
SWE
DENMARK
UNITED KINGDOM
IRELAND
NETH
E GER
BEL
W GER
FRANCE
S

UNITED STATES
OF AMERICA

PORTUGAL SPAIN

ITALY

MOROCCO

TUNISIA

ALGERIA

LI

WESTERN SAHARA

BERMUDA

ATLANTIC
OCEAN

MEXICO

MAURITANIA

MALI

NIGER

BAHAMAS

CUBA

PUERTO RICO (US)

DOM REP
HAITI

BELIZE
HONDURAS

JAMAICA

GUATEMALA
EL SALVADOR

NICARAGUA

VIRGIN ISLANDS (US)

GUADELOUPE (Fr)

DOMINICA

MARTINIQUE (Fr)

BARBADOS

TRINIDAD & TOBAGO

GAMBIA SENEGAL

GUINEA
BISSAU GUINEA

SIERRA-LEONE

BURKINA

IVORY
COAST GHANA

BENIN

NIGERIA

LIBERIA

COSTA RICA
PANAMA

VENEZUELA

GUYANA
SURINAME
FRENCH GUIANA (Fr)

ATLANTIC
OCEAN

CAMEROON

EQUATORIAL GUINEA

SAO TOME & PRINCIPE

GABON

CONGO

PACIFIC
OCEAN

COLOMBIA

ECUADOR

PERU

BRAZIL

BOLIVIA

PARAGUAY

WETLANDS LOSS

57	West Germany
55	Netherlands
42	Malaysia
29	Brunei
22	Finland
12	USA
9	Sweden

Proportion of wetlands
destroyed since 1950
percentages

CHILE

URUGUAY

ARGENTINA

FALKLAND ISLANDS (Br)

COASTAL CRISIS

DISTRIBUTION OF MAJOR
MANGROVE WETLANDS
mid-1980s sq kms

- 5000
- 1000
- 100

size of mangrove wetlands
unknown

countries with other
major wetlands

other countries

THREATS TO MANGROVE
WETLANDS

urban and agricultural
encroachment

mining and logging

pollution, oil spills, waste
dumping

Data compiled by Michael Steinitz   Main sources: Saenger (1983); AMBIO (1988).

The world's wetlands are being drained and paved over for development. More than half have already been destroyed.

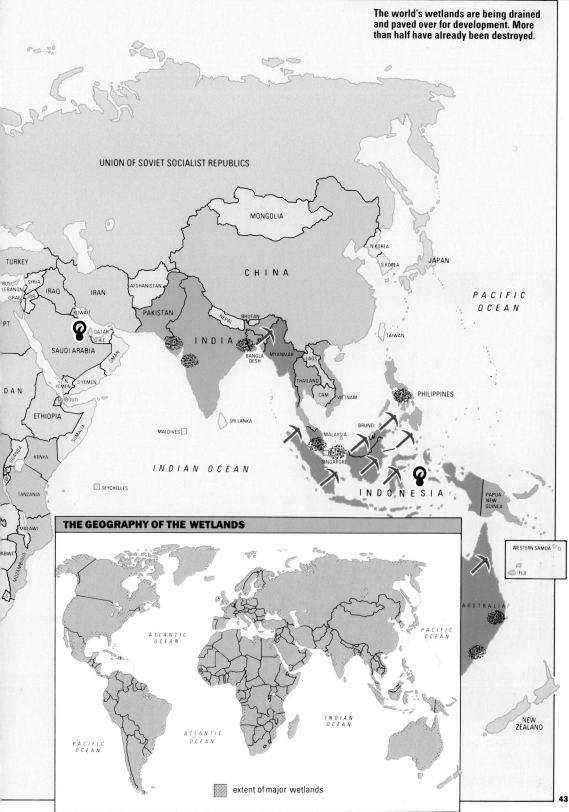

UNION OF SOVIET SOCIALIST REPUBLICS

MONGOLIA

N KOREA
S KOREA
JAPAN

TURKEY

RUSS SYRIA
LEBANON
ISRAEL JOR
IRAQ
IRAN
AFGHANISTAN

CHINA

PACIFIC
OCEAN

PT
KUWAIT
QATAR
UAE
PAKISTAN
NEPAL
BHUTAN

INDIA

TAIWAN

SAUDI ARABIA
OMAN
BANGLA
DESH
MYANMAR

DAN
N YEMEN
S YEMEN
DJIBOUTI

LAOS

THAILAND

ETHIOPIA

CAM
VIETNAM

PHILIPPINES

SRI LANKA

MALDIVES

BRUNEI

MALAYSIA

UGANDA
KENYA

SINGAPORE

INDIAN OCEAN

SEYCHELLES

TANZANIA

INDONESIA

PAPUA
NEW
GUINEA

MALAWI

ABWE

MOZAMBIQUE

THE GEOGRAPHY OF THE WETLANDS

ATLANTIC
OCEAN

PACIFIC
OCEAN

WESTERN SAMOA

FIJI

AUSTRALIA

PACIFIC
OCEAN

ATLANTIC
OCEAN

INDIAN
OCEAN

NEW
ZEALAND

PACIFIC
OCEAN

ATLANTIC
OCEAN

extent of major wetlands

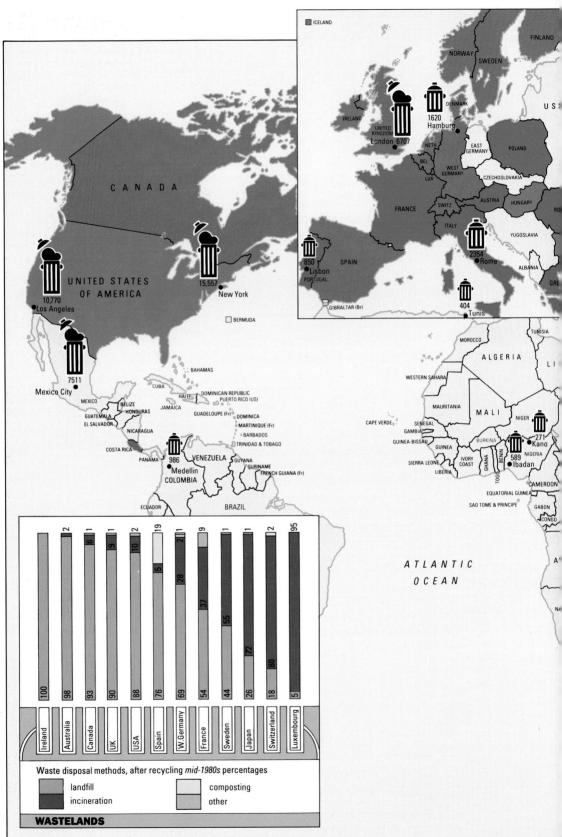

ICELAND

FINLAND

NORWAY SWEDEN

IRELAND

DENMARK

1620
Hamburg

London 6707

UNITED
KINGDOM

NETH.

EAST
GERMANY

POLAND

BEL
LUX

WEST
GERMANY

CZECHOSLOVAKIA

FRANCE

SWITZ

AUSTRIA

HUNGARY

ITALY

YUGOSLAVIA

850
Lisbon

SPAIN

2354
Rome

ALBANIA

PORTUGAL

GIBRALTAR (Br)

404
Tunis

U S

GRE

C A N A D A

UNITED STATES
OF AMERICA

10,770
Los Angeles

15,557
New York

7511
Mexico City

BAHAMAS

MEXICO

BELIZE

GUATEMALA
EL SALVADOR

HONDURAS

CUBA

HAITI

JAMAICA

DOMINICAN REPUBLIC
PUERTO RICO (US)

GUADELOUPE (Fr)

DOMINICA

MARTINIQUE (Fr)

BARBADOS

TRINIDAD & TOBAGO

BERMUDA

NICARAGUA

COSTA RICA

PANAMA

986
Medellin

COLOMBIA

ECUADOR

VENEZUELA

GUYANA

SURINAME

FRENCH GUIANA (Fr)

BRAZIL

TUNISIA

MOROCCO

ALGERIA

LI

WESTERN SAHARA

MAURITANIA

MALI

NIGER

CAPE VERDE

SENEGAL

GAMBIA

GUINEA-BISSAU

GUINEA

SIERRA LEONE

LIBERIA

IVORY
COAST

GHANA

BURKINA

BENIN
TOGO

271
Kano

589
Ibadan

NIGERIA

CAMEROON

EQUATORIAL GUINEA

SAO TOME & PRINCIPE

GABON

CONGO

A

ATLANTIC
OCEAN

NA

Waste disposal methods, after recycling *mid-1980s* percentages

	Ireland	Australia	Canada	UK	USA	Spain	W.Germany	France	Sweden	Japan	Switzerland	Luxembourg
top		2	1	1	2	19	1	9	1	1	2	95
			6	9	10	5	2	37	55	72	80	5
	100	98	93	90	88	76	28	54	44	26	18	
							69					

landfill

composting

incineration

other

WASTELANDS

Data compiled by Joni Seager   Main sources: Cointreau (1987); Levenson, US Congress Office of Technology Assessment; OECD *Environmental Data Compendium* (1989).

The richer the country the more garbage it produces. But getting rid of it safely is a global problem.

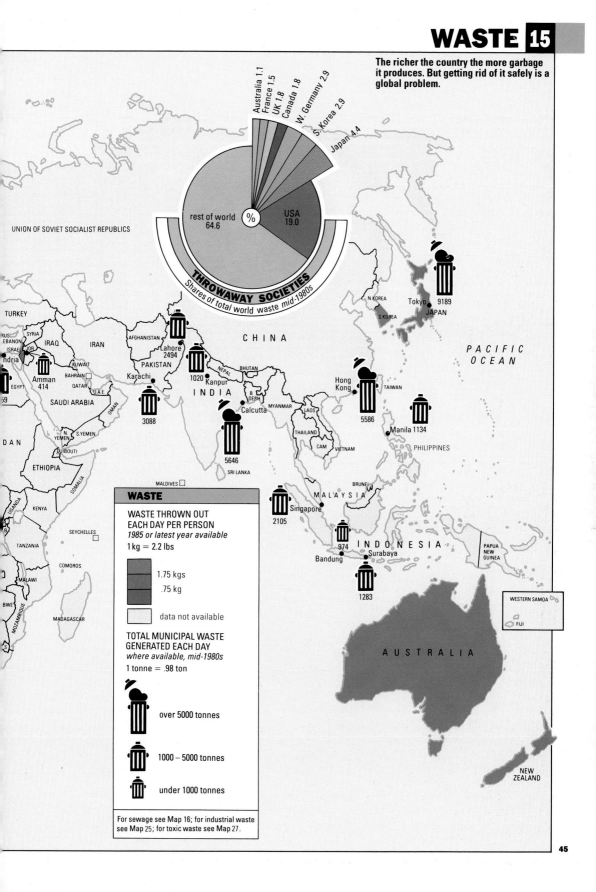

Australia 1.1
France 1.5
UK 1.8
Canada 1.8
W. Germany 2.9
S. Korea 2.9
Japan 4.4

rest of world 64.6

%

USA 19.0

THROWAWAY SOCIETIES
Shares of total world waste mid-1980s

UNION OF SOVIET SOCIALIST REPUBLICS

CHINA

N KOREA

S KOREA

Tokyo 9189
JAPAN

PACIFIC OCEAN

TURKEY

SYRIA
LEBANON
ISRAEL JOR
India

IRAQ

IRAN

AFGHANISTAN

Lahore 2494

PAKISTAN

Karachi

NEPAL

BHUTAN

B. DESH

INDIA

Kanpur 1020

Calcutta

MYANMAR

LAOS

THAILAND

CAM
VIETNAM

Hong Kong

TAIWAN

5586

Manila 1134

PHILIPPINES

KUWAIT

BAHRAIN

QATAR
U.A.E

Amman 414

SAUDI ARABIA

OMAN

EGYPT

3088

N.
YEMEN

S.YEMEN

D.JIBOUTI

DAN

ETHIOPIA

SOMALIA

5646

SRI LANKA

MALDIVES

BRUNEI

MALAYSIA

Singapore

2105

INDONESIA

PAPUA
NEW
GUINEA

UGANDA

KENYA

SEYCHELLES

TANZANIA

COMOROS

MALAWI

MADAGASCAR

BWE

MOZAMBIQUE

974

Bandung

Surabaya

1283

WESTERN SAMOA

FIJI

AUSTRALIA

NEW
ZEALAND

WASTE

**WASTE THROWN OUT
EACH DAY PER PERSON**
1985 or latest year available
1 kg = 2.2 lbs

1.75 kgs
.75 kg

data not available

**TOTAL MUNICIPAL WASTE
GENERATED EACH DAY**
where available, mid-1980s
1 tonne = .98 ton

over 5000 tonnes

1000 – 5000 tonnes

under 1000 tonnes

For sewage see Map 16; for industrial waste see Map 25; for toxic waste see Map 27.

45

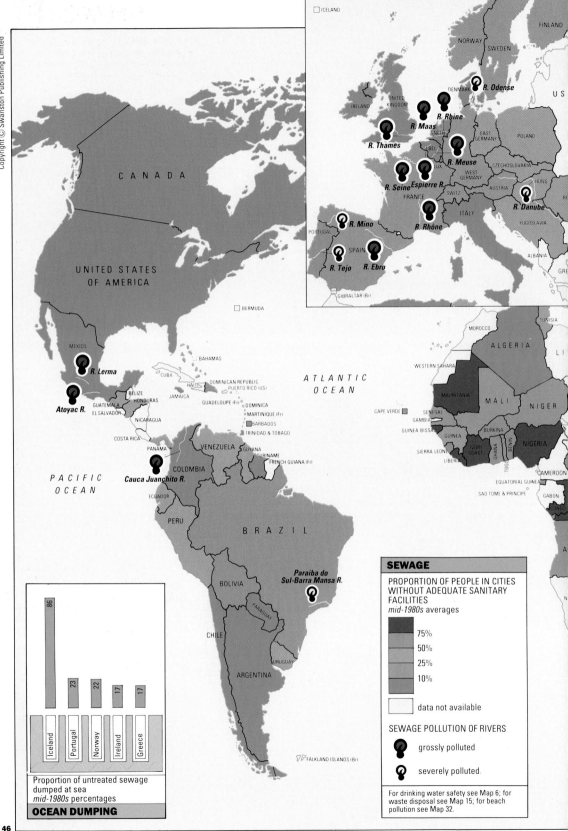

Data compiled by Lesley Downing   Main sources: Monitoring and Assessment Research Centre (1987); UNEP (1989); WHO *World Health Statistics Annual (1986).*

The growth of city populations is overwhelming existing systems for safe disposal of sewage. Many urban rivers have become virtual open sewers.

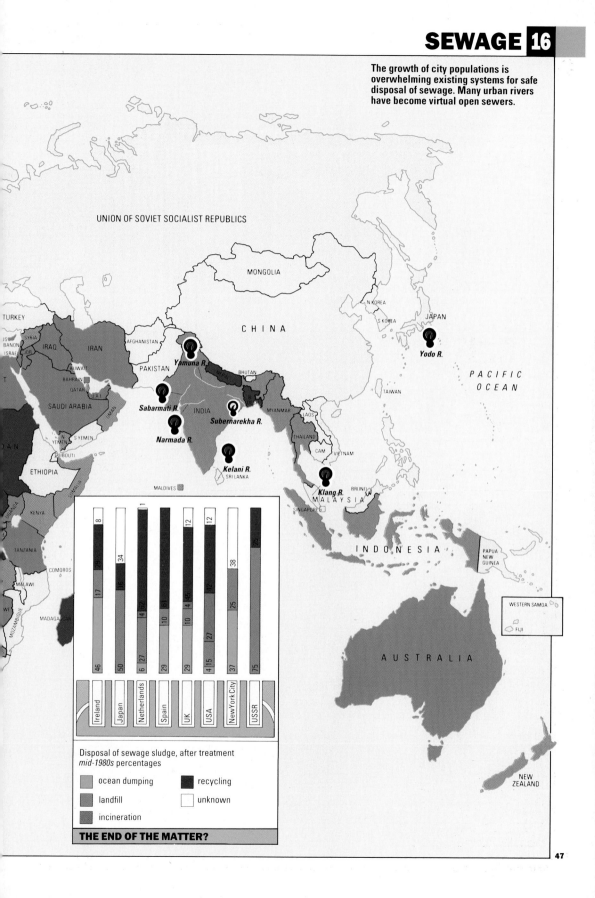

UNION OF SOVIET SOCIALIST REPUBLICS

MONGOLIA

CHINA

TURKEY

IRAQ   IRAN

AFGHANISTAN

PAKISTAN

BHUTAN

Yamuna R.

SAUDI ARABIA

INDIA

Sabarmati R.

MYANMAR

Subernarekha R.

Narmada R.

THAILAND

Kelani R.
SRI LANKA

MALDIVES

Klang R.
MALAYSIA

SINGAPORE

N KOREA

S KOREA

JAPAN

Yodo R.

PACIFIC
OCEAN

TAIWAN

LAOS

CAM   VIETNAM

BRUNEI

ETHIOPIA

KENYA

TANZANIA

COMOROS

MALAWI

MADAGASCAR

MOZAMBIQUE

INDONESIA

PAPUA
NEW
GUINEA

WESTERN SAMOA

FIJI

AUSTRALIA

NEW
ZEALAND

Disposal of sewage sludge, after treatment
mid-1980s percentages

- ocean dumping
- landfill
- incineration
- recycling
- unknown

THE END OF THE MATTER?

Chart values by country:

	Ireland	Japan	Netherlands	Spain	UK	USA	New York City	USSR
(top)	8	34	1	12	12		38	25
	29	16	62	61	46	47	25	
	17		4	10	4	27		
		50	6|27	29	10	4|15	37	75
(bottom)	46				29			

AIR QUALITY

CITIES WITH AIR QUALITY PROBLEMS
Annual averages *mid-1980s*
Known pollution is greatest from high levels of

- sulphur dioxide
- nitrogen oxides
- both

High level exceeds international safe health guidelines

medium levels of:

- sulphur dioxide
- nitrogen oxides
- both

Most sulphur dioxide comes from burning fossil fuels; most nitrogen oxides come from cars.

For acid rain see Map 24; for fossil fuel pollution see Map 23.

Data compiled by Claire Holman Main sources: GEMS (1988); World Resources Institute (1988); OECD (1989).

In many cities, just breathing is a health hazard.

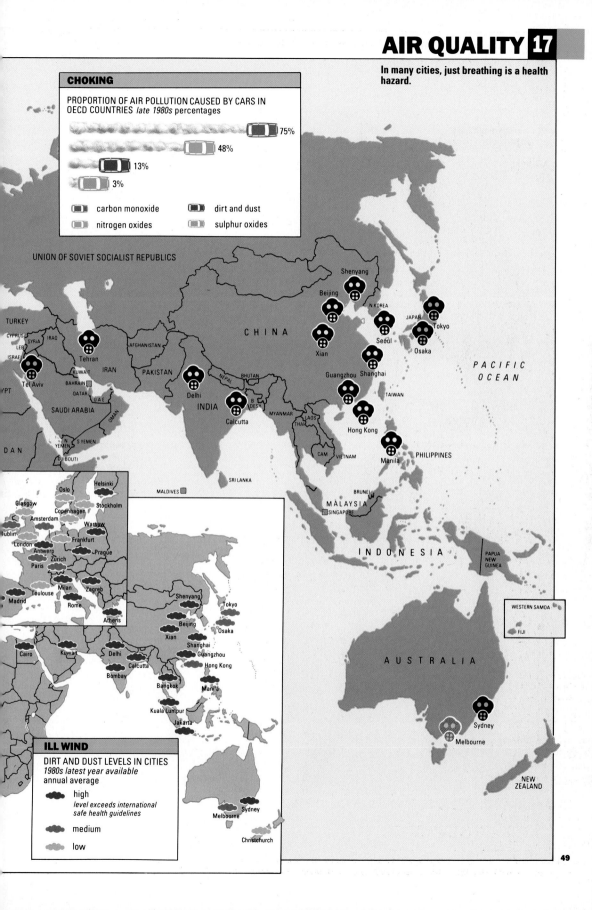

CHOKING

PROPORTION OF AIR POLLUTION CAUSED BY CARS IN OECD COUNTRIES *late 1980s* percentages

- 75%
- 48%
- 13%
- 3%

- carbon monoxide
- nitrogen oxides
- dirt and dust
- sulphur oxides

UNION OF SOVIET SOCIALIST REPUBLICS

TURKEY
CYPRUS
SYRIA IRAQ
LEB
ISRAEL
Tel Aviv
YPT
KUWAIT
BAHRAIN
QATAR
U A E
SAUDI ARABIA
N S YEMEN
YEMEN
DJIBOUTI
DAN

Tehran
IRAN
AFGHANISTAN
PAKISTAN
NEPAL
BHUTAN
Delhi
INDIA
B DESH
Calcutta
MYANMAR
THAI
LAOS
CAM
VIETNAM
SRI LANKA
MALDIVES

CHINA
Shenyang
Beijing
N KOREA
Xian
Guangzhou Shanghai
Seoul
JAPAN
Tokyo
Osaka
TAIWAN
Hong Kong
Manila
PHILIPPINES

PACIFIC OCEAN

BRUNEI
MALAYSIA
SINGAPORE
INDONESIA
PAPUA NEW GUINEA

WESTERN SAMOA
FIJI

AUSTRALIA

Sydney
Melbourne

NEW ZEALAND

Helsinki
Oslo
Glasgow
Copenhagen
Stockholm
Amsterdam
Warsaw
Dublin
London
Frankfurt
Antwerp
Prague
Paris
Zürich
Milan
Zagreb
Madrid
Toulouse
Rome
Athens
Shenyang
Tokyo
Beijing
Osaka
Xian
Shanghai
Cairo
Kuwait
Delhi
Guangzhou
Calcutta
Hong Kong
Bombay
Bangkok
Manila
Kuala Lumpur
Jakarta

Melbourne
Sydney
Christchurch

ILL WIND

DIRT AND DUST LEVELS IN CITIES
1980s latest year available
annual average

- high
  level exceeds international safe health guidelines
- medium
- low

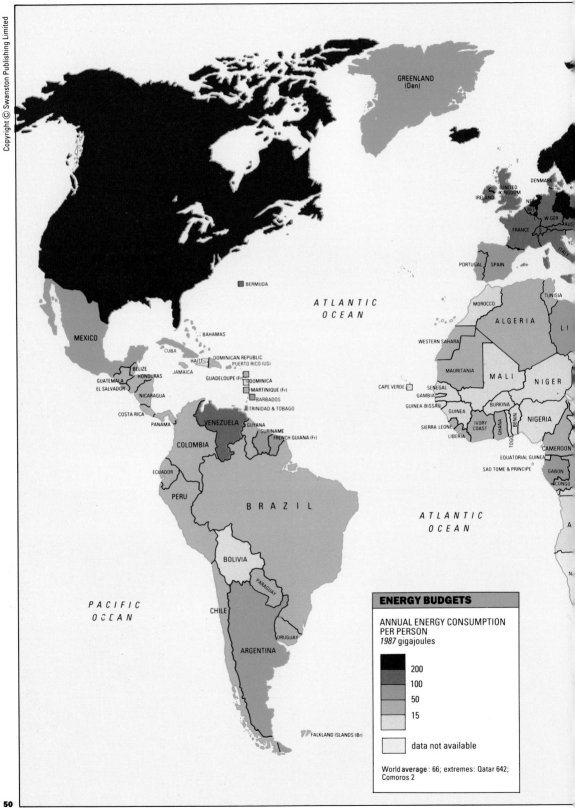

Data compiled by Lisa Greber   Main sources: IEA (1989); UN *Energy Statistics Yearbook* (1989).

There is neither social nor environmental balance in today's global energy budget.

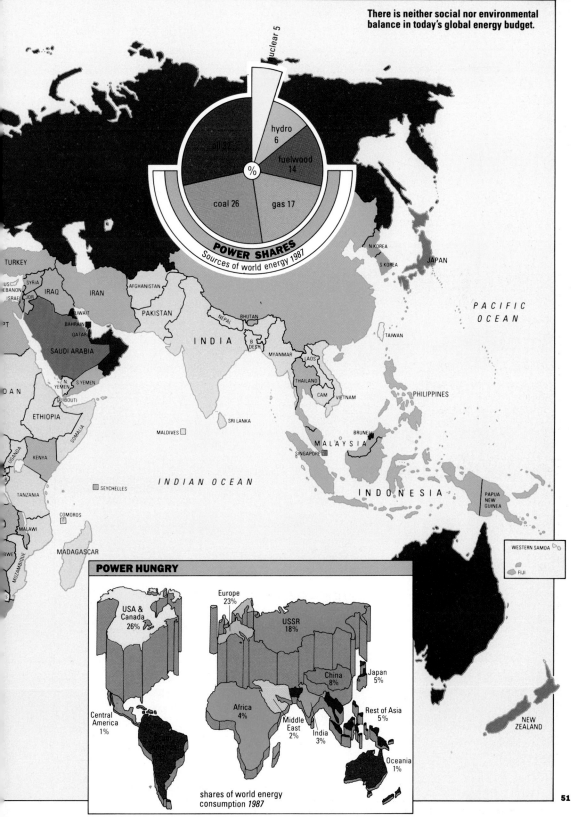

POWER SHARES
Sources of world energy 1987

%

- nuclear 5
- hydro 6
- fuelwood 14
- oil 32.5
- gas 17
- coal 26

POWER HUNGRY

USA & Canada 26%

Europe 23%

USSR 18%

China 8%

Japan 5%

Central America 1%

Africa 4%

Middle East 2%

India 3%

Rest of Asia 5%

Oceania 1%

shares of world energy consumption 1987

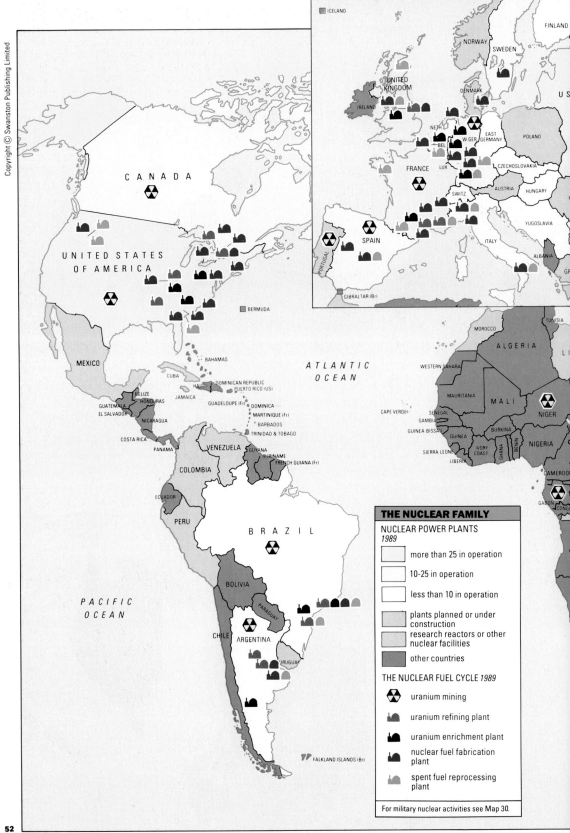

THE NUCLEAR FAMILY

NUCLEAR POWER PLANTS
1989

more than 25 in operation

10-25 in operation

less than 10 in operation

plants planned or under construction

research reactors or other nuclear facilities

other countries

THE NUCLEAR FUEL CYCLE *1989*

uranium mining

uranium refining plant

uranium enrichment plant

nuclear fuel fabrication plant

spent fuel reprocessing plant

For military nuclear activities see Map 30.

Data compiled by Lisa Greber   Main sources: International Atomic Energy Agency (1989); *Nuclear News* (August 1989).

There are more than 400 nuclear power plants operating in 26 countries. But an even greater number of states participate in the supply and support of the nuclear industry.

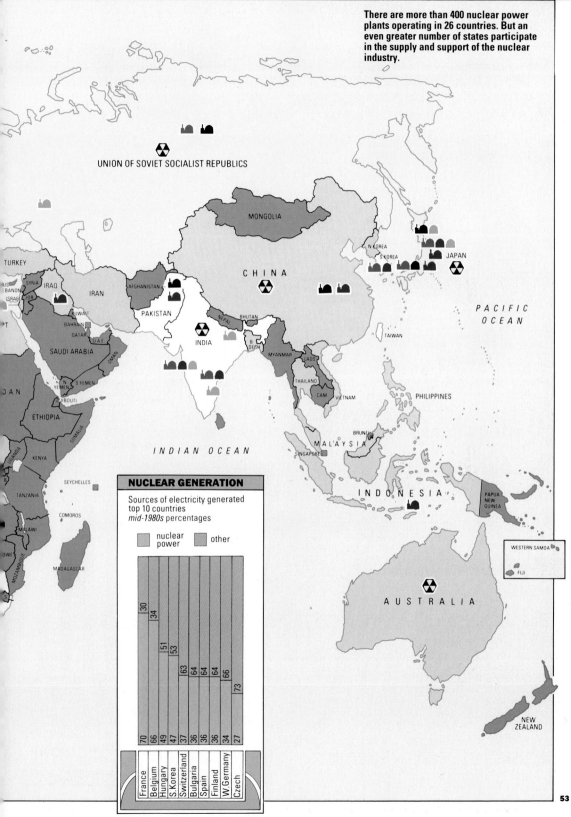

UNION OF SOVIET SOCIALIST REPUBLICS

MONGOLIA

N KOREA

S KOREA

JAPAN

CHINA

PACIFIC OCEAN

TURKEY

SYRIA

IRAQ

IRAN

AFGHANISTAN

PAKISTAN

NEPAL

BHUTAN

TAIWAN

INDIA

B.DESH

MYANMAR

LAOS

THAILAND

CAM

VIETNAM

PHILIPPINES

SAUDI ARABIA

OMAN

N YEMEN

S YEMEN

DJIBOUTI

ETHIOPIA

KENYA

MALAYSIA

BRUNEI

SINGAPORE

INDIAN OCEAN

INDONESIA

PAPUA NEW GUINEA

SEYCHELLES

TANZANIA

COMOROS

MALAWI

MOZAMBIQUE

MADAGASCAR

WESTERN SAMOA

FIJI

AUSTRALIA

NEW ZEALAND

NUCLEAR GENERATION

Sources of electricity generated
top 10 countries
mid-1980s percentages

nuclear power   other

	France	Belgium	Hungary	S.Korea	Switzerland	Bulgaria	Spain	Finland	W.Germany	Czech
other									30	34
			51	53						
					63	64	64	66		73
nuclear	70	66	49	47	37	36	36	36	34	27

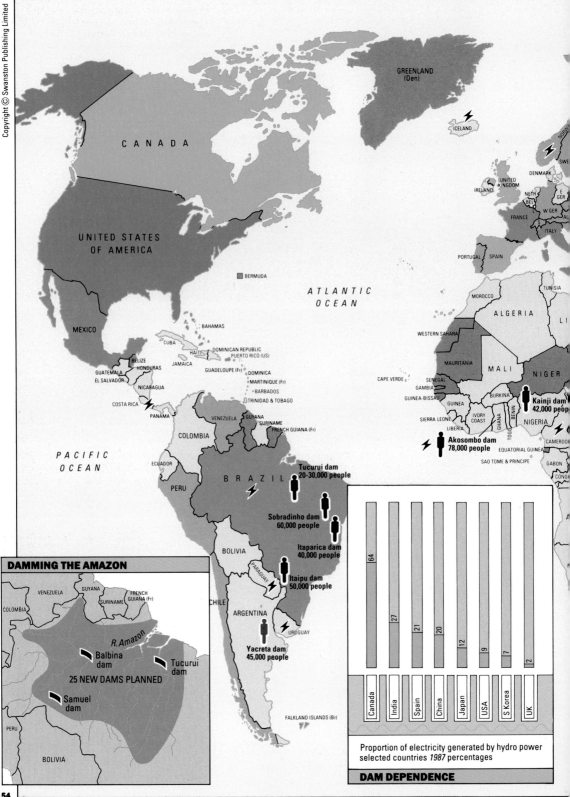

CANADA

GREENLAND
(Den)

ICELAND

UNITED STATES
OF AMERICA

BERMUDA

ATLANTIC
OCEAN

MEXICO

BAHAMAS

CUBA
HAITI
DOMINICAN REPUBLIC
PUERTO RICO (US)
JAMAICA
BELIZE
HONDURAS
GUATEMALA
EL SALVADOR
NICARAGUA
GUADELOUPE (Fr) DOMINICA
MARTINIQUE (Fr)
BARBADOS
TRINIDAD & TOBAGO
COSTA RICA
PANAMA

DENMARK
UNITED KINGDOM
IRELAND
NETH
BEL
E GER
W GER
FRANCE
S
ITALY
PORTUGAL SPAIN
NORW
SWE

AL

TUNISIA
MOROCCO
ALGERIA
LI
WESTERN SAHARA
MAURITANIA
MALI
NIGER
CAPE VERDE
SENEGAL
GAMBIA
GUINEA-BISSAU GUINEA
SIERRA LEONE
IVORY
COAST
LIBERIA
BURKINA
GHANA
TOGO
BENIN
NIGERIA
EQUATORIAL GUINEA
SAO TOME & PRINCIPE
CAMEROO
GABON
CONG

Kainji dam
42,000 peop

Akosombo dam
78,000 people

PACIFIC
OCEAN

COLOMBIA
VENEZUELA GUYANA
SURINAME
FRENCH GUIANA (Fr)
ECUADOR
PERU
B R A Z I L

Tucurui dam
20-30,000 people

Sobradinho dam
60,000 people

Itaparica dam
40,000 people

BOLIVIA
PARAGUAY
Itaipu dam
50,000 people

CHILE
ARGENTINA
URUGUAY

Yacreta dam
45,000 people

FALKLAND ISLANDS (Br)

DAMMING THE AMAZON

VENEZUELA GUYANA
COLOMBIA SURINAME
FRENCH
GUIANA (Fr)

R. Amazon

Balbina
dam
Tucurui
dam

25 NEW DAMS PLANNED

Samuel
dam

PERU

BOLIVIA

Canada	India	Spain	China	Japan	USA	S Korea	UK
64	27	21	20	12	9	7	2

Proportion of electricity generated by hydro power
selected countries *1987* percentages

DAM DEPENDENCE

Data compiled by Constance E Hunt   Main sources: International Commission on Large Dams (1984); Mermel (1989); Simons (1989);

Hydro power is a cheap and renewable energy source, but dams often carry a high social and environmental price tag.

UNION OF SOVIET SOCIALIST REPUBLICS

MONGOLIA

N.KOREA

S.KOREA

JAPAN

CHINA

TURKEY

Keban dam 30,000 people

RUS
LEBANON
SYRIA
ISRAEL
JOR
IRAQ
IRAN
AFGHANISTAN
KUWAIT
BAHRAIN
QATAR
U.A.E.
SAUDI ARABIA
PAKISTAN
NEPAL
BHUTAN
INDIA
B.DESH
MYANMAR
S.YEMEN
YEMEN
DJIBOUTI

PT

Aswan dam 20,000 people

DAN

ETHIOPIA
SOMALIA
KENYA

N.
S.YEMEN

Sardar Sarovar dam 70,000 people

Three Gorges dam 1 million people

TAIWAN

PACIFIC OCEAN

THAILAND
LAOS
CAM
VIETNAM
PHILIPPINES

SRI LANKA

MALDIVES

INDIAN OCEAN

SEYCHELLES

BRUNEI
MALAYSIA
SINGAPORE

INDONESIA

PAPUA NEW GUINEA

BURUNDI
TANZANIA

COMOROS

MADAGASCAR

ZBIA
MALAWI
MOZAMBIQUE
ZBWE

Cabora Bassa dam 25,000 people

S

Kariba dam 57,000 people

WESTERN SAMOA

FIJI

AUSTRALIA

NEW ZEALAND

HYDRO POWER

NUMBER OF LARGE DAMS
Dams over 15 metres high *mid-1980s*

- 1000
- 500
- 250
- 50

data not available

DAMS WITH EXCEPTIONAL IMPACT ON THE ENVIRONMENT

dam known to have forced relocation of more than 10,000 people *since early 1970s*

planned dam which would force major relocation of people

DAMMED ELECTRICITY

over 90% of electricity generated by hydro power

Extreme: China 18,595 dams

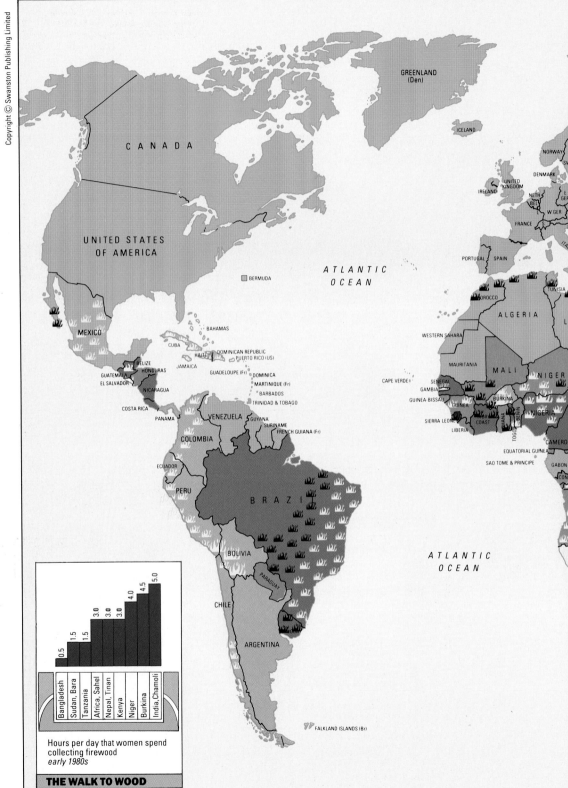

Hours per day that women spend
collecting firewood
early 1980s

THE WALK TO WOOD

Data compiled by Thomas E Downing   Main sources: Agarwal (1986); FAO (1981); UN Statistical Office (1987).

Firewood, the primary source of energy for many people in the world, is increasingly short in supply.

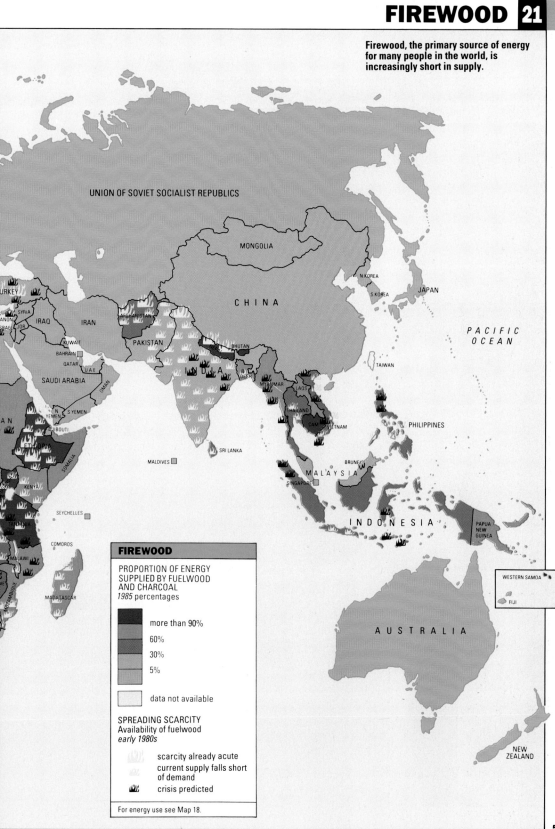

UNION OF SOVIET SOCIALIST REPUBLICS

MONGOLIA

N.KOREA

S.KOREA

JAPAN

CHINA

PACIFIC OCEAN

TURKEY

US...
LEBANON
SYRIA
ISRAEL JOR.
IRAQ

IRAN

AFGHANISTAN

PAKISTAN

BHUTAN

TAIWAN

KUWAIT
BAHRAIN
QATAR
U.A.E

SAUDI ARABIA

OMAN

N.
YEMEN
S.YEMEN

DJIBOUTI

ETHIOPIA

SOMALIA

KENYA

SEYCHELLES

TANZANIA

COMOROS

MALAWI

MADAGASCAR

MALDIVES

SRI LANKA

I N D I A

B.
DESH

MYANMAR

LAOS

THAILAND

CAM.

VIETNAM

B.

PHILIPPINES

BRUNEI

MALAYSIA

SINGAPORE

I N D O N E S I A

PAPUA
NEW
GUINEA

WESTERN SAMOA

FIJI

A U S T R A L I A

NEW
ZEALAND

FIREWOOD

PROPORTION OF ENERGY
SUPPLIED BY FUELWOOD
AND CHARCOAL
1985 percentages

- more than 90%
- 60%
- 30%
- 5%
- data not available

SPREADING SCARCITY
Availability of fuelwood
early 1980s

- scarcity already acute
- current supply falls short of demand
- crisis predicted

For energy use see Map 18.

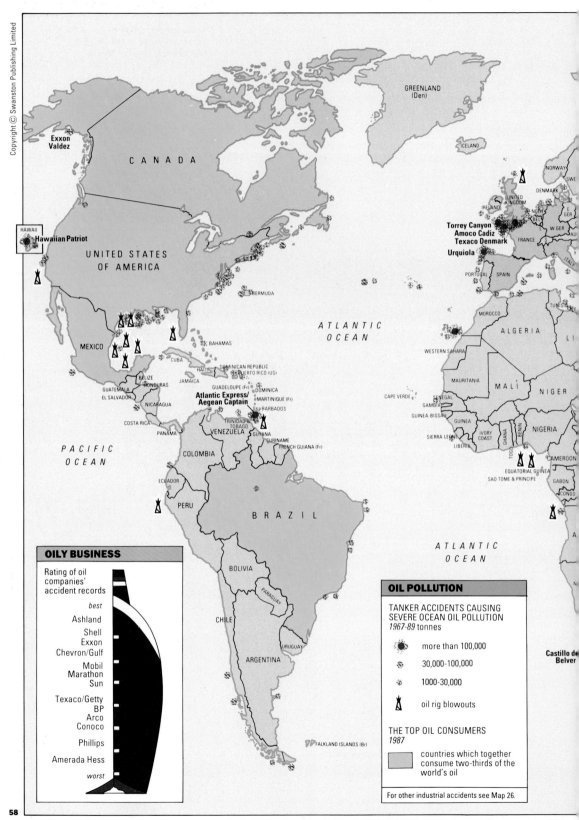

Copyright © Swanston Publishing Limited

GREENLAND (Den)

ICELAND

NORWAY
SWE
DENMARK
E. GER
NETH
BEL
W. GER
FRANCE
ITAL

Exxon
Valdez

C A N A D A

UNITED KINGDOM
IRELAND

Torrey Canyon
Amoco Cadiz
Texaco Denmark
Urquiola

HAWAII
Hawaiian Patriot

UNITED STATES
OF AMERICA

PORTUGAL   SPAIN

MOROCCO

TUNISIA

ALGERIA

LI

BERMUDA

ATLANTIC
OCEAN

WESTERN SAHARA

MEXICO

BAHAMAS

CUBA

HAITI   DOMINICAN REPUBLIC
          PUERTO RICO (US)
GUADELOUPE (Fr)

DOMINICA
MARTINIQUE (Fr)

ST LUCIA
BARBADOS

MAURITANIA

M A L I

N I G E R

BELIZE
GUATEMALA   HONDURAS
EL SALVADOR
NICARAGUA
COSTA RICA

JAMAICA

CAPE VERDE

SENEGAL
GAMBIA
GUINEA-BISSAU

Atlantic Express/
Aegean Captain

GUINEA

SIERRA LEONE

IVORY
COAST

GHANA
TOGO
BENIN

NIGERIA

PANAMA

TRINIDAD &
TOBAGO
VENEZUELA   GUYANA
          SURINAME
          FRENCH GUIANA (Fr)

LIBERIA

CAMEROON

PACIFIC
OCEAN

COLOMBIA

EQUATORIAL GUINEA
SAO TOME & PRINCIPE

GABON
CONGO

ECUADOR

PERU

B R A Z I L

ATLANTIC
OCEAN

BOLIVIA

PARAGUAY

CHILE

URUGUAY

ARGENTINA

Castillo de
Belver

FALKLAND ISLANDS (Br)

OIL POLLUTION

TANKER ACCIDENTS CAUSING
SEVERE OCEAN OIL POLLUTION
1967-89 tonnes

more than 100,000

30,000-100,000

1000-30,000

oil rig blowouts

THE TOP OIL CONSUMERS
1987

countries which together
consume two-thirds of the
world's oil

For other industrial accidents see Map 26.

Data compiled by Dominic Golding   Main sources: Couper (1983); ITOPF (1983); Levy (1984); McKenzie (1989); NAS (1985); OECD *Environmental Data Compendium* (1989); Tanker Advisory Center (1989); UNEP *Environmental Data Report* (1989).

Dependence on oil is a defining feature of modern industrial economies, as is the pollution it leaves in its wake.

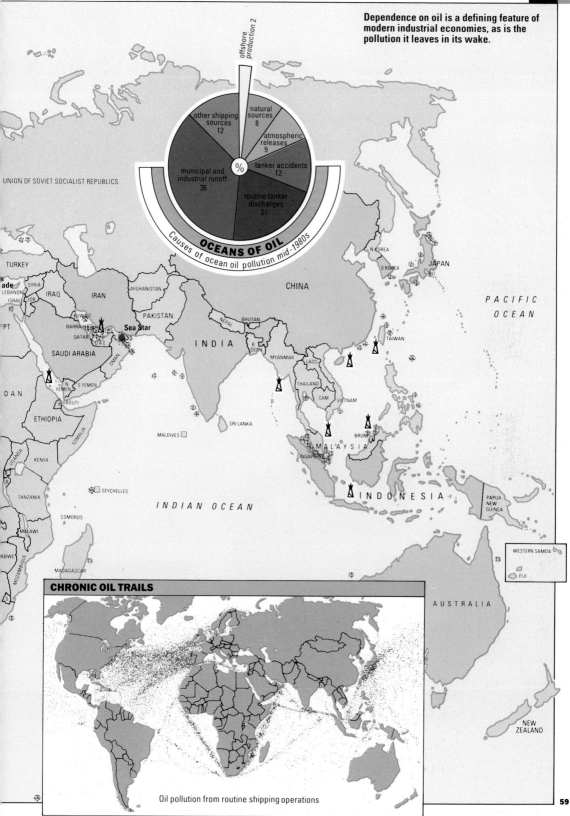

OCEANS OF OIL
Causes of ocean oil pollution mid-1980s

offshore production 2

natural sources 8

atmospheric releases 9

tanker accidents 12

other shipping sources 12

municipal and industrial runoff 36

routine tanker discharges 21

UNION OF SOVIET SOCIALIST REPUBLICS

TURKEY
ade
SYRIA
LEBANON
JOR
ISRAEL
IRAQ
IRAN
AFGHANISTAN
KUWAIT
PAKISTAN
BAHRAIN
QATAR
UAE
OMAN
Sea Star
SAUDI ARABIA
N YEMEN
S YEMEN
DJIBOUTI
DAN
ETHIOPIA
SOMALIA
UGANDA
KENYA
TANZANIA
COMOROS
MALAWI
ABWE
MOZAMBIQUE
MADAGASCAR
SEYCHELLES
MALDIVES
SRI LANKA
INDIA
NEPAL
BHUTAN
B DESH
MYANMAR
LAOS
THAILAND
CAM
VIETNAM
MALAYSIA
SINGAPORE
BRUNEI
INDONESIA
CHINA
N KOREA
S KOREA
JAPAN
TAIWAN
PAPUA NEW GUINEA
AUSTRALIA
WESTERN SAMOA
FIJI
NEW ZEALAND

PACIFIC OCEAN

INDIAN OCEAN

CHRONIC OIL TRAILS

Oil pollution from routine shipping operations

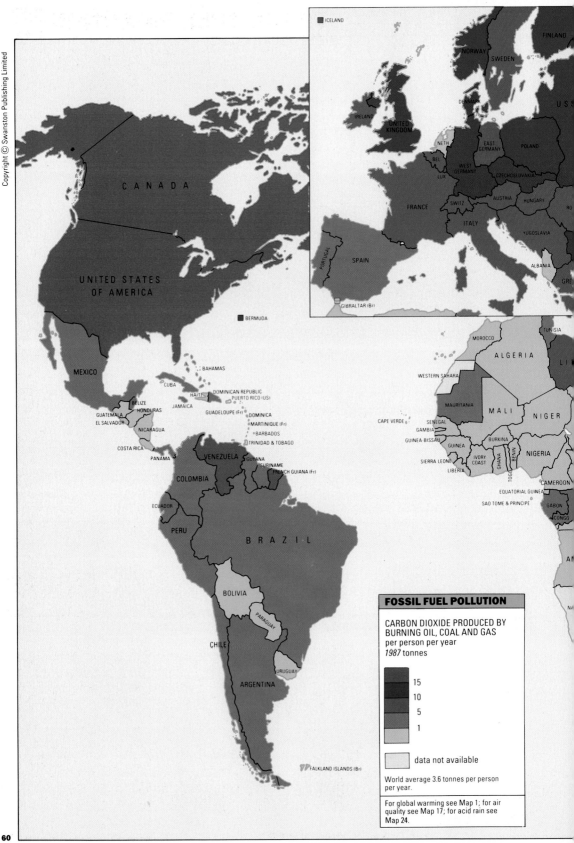

ICELAND

FINLAND

NORWAY

SWEDEN

DENMARK

USS

IRELAND

UNITED
KINGDOM

NETH

EAST
GERMANY

POLAND

BEL

LUX

WEST
GERMANY

CZECHOSLOVAKIA

FRANCE

SWITZ

AUSTRIA

HUNGARY

RO

ITALY

YUGOSLAVIA

PORTUGAL

SPAIN

ALBANIA

GRE

GIBRALTAR (Br)

C A N A D A

UNITED STATES
OF AMERICA

BERMUDA

TUNISIA

MOROCCO

ALGERIA

LI

MEXICO

WESTERN SAHARA

BAHAMAS

CUBA

DOMINICAN REPUBLIC
PUERTO RICO (US)

MAURITANIA

MALI

NIGER

HAITI

BELIZE
HONDURAS

JAMAICA

CAPE VERDE

GUATEMALA
EL SALVADOR

GUADELOUPE (Fr)

DOMINICA

SENEGAL

NICARAGUA

MARTINIQUE (Fr)

GAMBIA

GUINEA-BISSAU

BURKINA

NIGERIA

COSTA RICA

BARBADOS

TRINIDAD & TOBAGO

GUINEA

PANAMA

VENEZUELA

GUYANA

SIERRA LEONE

IVORY
COAST

GHANA

BENIN

LIBERIA

TOGO

SURINAME

FRENCH GUIANA (Fr)

CAMEROON

COLOMBIA

EQUATORIAL GUINEA

ECUADOR

SAO TOME & PRINCIPE

GABON

CONGO

PERU

B R A Z I L

AM

BOLIVIA

NA

PARAGUAY

FOSSIL FUEL POLLUTION

CHILE

CARBON DIOXIDE PRODUCED BY
BURNING OIL, COAL AND GAS
per person per year
1987 tonnes

URUGUAY

ARGENTINA

15

10

5

1

data not available

World average 3.6 tonnes per person
per year.

FALKLAND ISLANDS (Br)

For global warming see Map 1; for air
quality see Map 17; for acid rain see
Map 24.

Data compiled by Claire Holman   Main sources: Eliasson et.al. (1988); GEMS (1988); Oak Ridge National Laboratory (1988).

There is now 25 per cent more carbon dioxide in the atmosphere than there was at the beginning of the Industrial Revolution. It is the major cause of global warming.

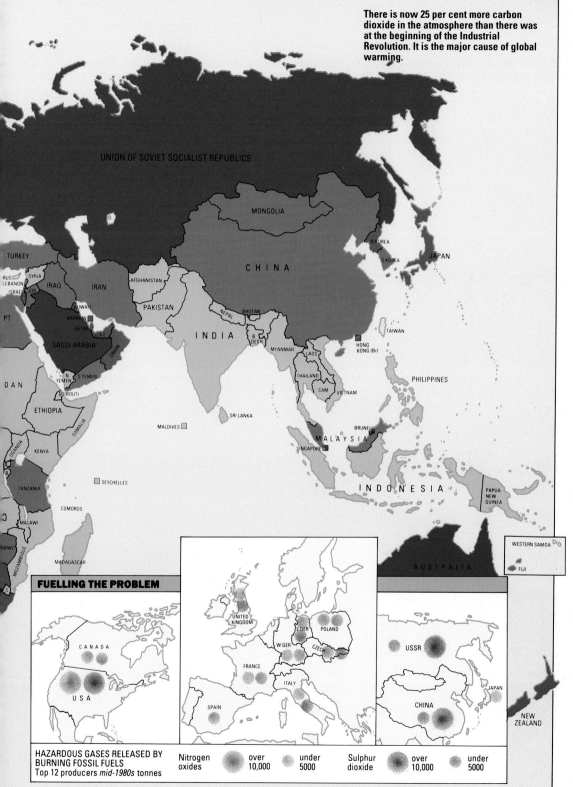

UNION OF SOVIET SOCIALIST REPUBLICS

MONGOLIA

TURKEY

RUS
LEBANON
ISRAE
SYRIA
JOR
IRAQ
IRAN
AFGHANISTAN

N KOREA
S KOREA
JAPAN

C H I N A

PT

KUWAIT
BAHRAIN
QATAR
U.A.E.
SAUDI ARABIA
OMAN

PAKISTAN

NEPAL
BHUTAN

TAIWAN

I N D I A

B
DESH
MYANMAR

HONG
KONG (Br)

DAN

N
YEMEN
S YEMEN
DJIBOUTI

LAOS

THAILAND

PHILIPPINES

ETHIOPIA

SOMALIA

SRI LANKA

CAM
VIETNAM

MALDIVES

UGANDA
KENYA

BRUNEI
M A L A Y S I A
SINGAPORE

TANZANIA

SEYCHELLES

I N D O N E S I A

PAPUA
NEW
GUINEA

COMOROS

MALAWI

MADAGASCAR

ABWE

MOZAMBIQUE

WESTERN SAMOA

FIJI

A U S T R A L I A

FUELLING THE PROBLEM

CANADA

U S A

UNITED
KINGDOM

E GER
POLAND

W GER
CZECH

FRANCE

ITALY

SPAIN

USSR

JAPAN

CHINA

NEW
ZEALAND

HAZARDOUS GASES RELEASED BY BURNING FOSSIL FUELS Top 12 producers *mid-1980s* tonnes	Nitrogen oxides	over 10,000	under 5000	Sulphur dioxide	over 10,000	under 5000

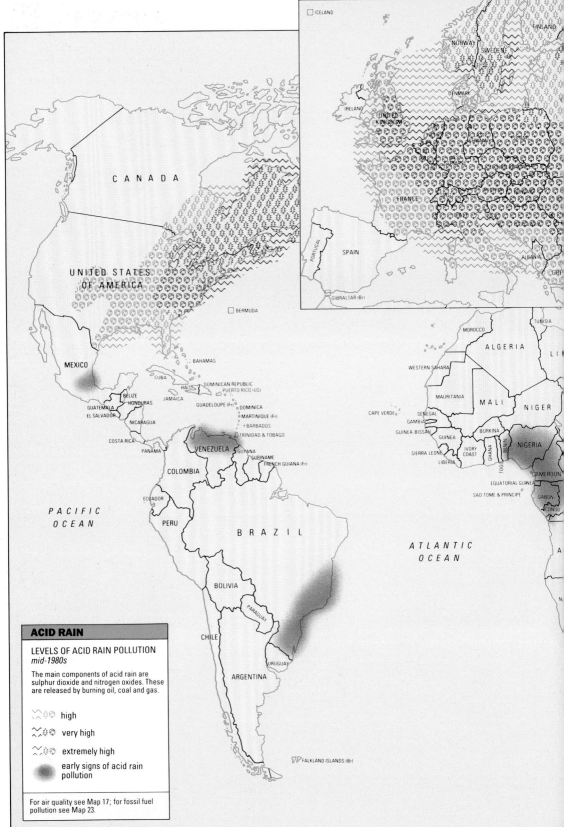

ACID RAIN

LEVELS OF ACID RAIN POLLUTION
mid-1980s

The main components of acid rain are
sulphur dioxide and nitrogen oxides. These
are released by burning oil, coal and gas.

高 high

very high

extremely high

early signs of acid rain
pollution

For air quality see Map 17; for fossil fuel
pollution see Map 23.

CANADA

UNITED STATES
OF AMERICA

BERMUDA

MEXICO

BAHAMAS

CUBA

BELIZE
GUATEMALA
EL SALVADOR
HONDURAS
NICARAGUA
COSTA RICA
PANAMA

HAITI
JAMAICA

DOMINICAN REPUBLIC
PUERTO RICO (US)
GUADELOUPE (Fr)
DOMINICA
MARTINIQUE (Fr)
BARBADOS
TRINIDAD & TOBAGO

VENEZUELA
GUYANA
SURINAME
FRENCH GUIANA (Fr)

COLOMBIA

ECUADOR

PERU

BRAZIL

BOLIVIA

PARAGUAY

CHILE

URUGUAY

ARGENTINA

PACIFIC
OCEAN

ATLANTIC
OCEAN

FALKLAND ISLANDS (Br)

ICELAND

NORWAY
SWEDEN
FINLAND

DENMARK

IRELAND

UNITED
KINGDOM

FRANCE

PORTUGAL

SPAIN

GIBRALTAR (Br)

ALBANIA

MOROCCO

TUNISIA

ALGERIA

WESTERN SAHARA

MAURITANIA

MALI

NIGER

CAPE VERDE

SENEGAL
GAMBIA
GUINEA BISSAU
GUINEA
SIERRA LEONE
LIBERIA

BURKINA

IVORY
COAST

GHANA

TOGO
BENIN

NIGERIA

CAMEROON

EQUATORIAL GUINEA

SAO TOME & PRINCIPE

GABON

CONGO

Data compiled by Friends of the Earth/Mike Birkin   Main sources: Goldsmith & Hildyard eds. (1988); UNEP, *Forest Damage & Air Pollution* (1989);
McCormick (1989); World Resources Institute (1988).

Acid rain is the fallout of industrial and fossil-fuel dependent economies. After falling on Europe for more than 100 years it is now becoming a global phenomenon.

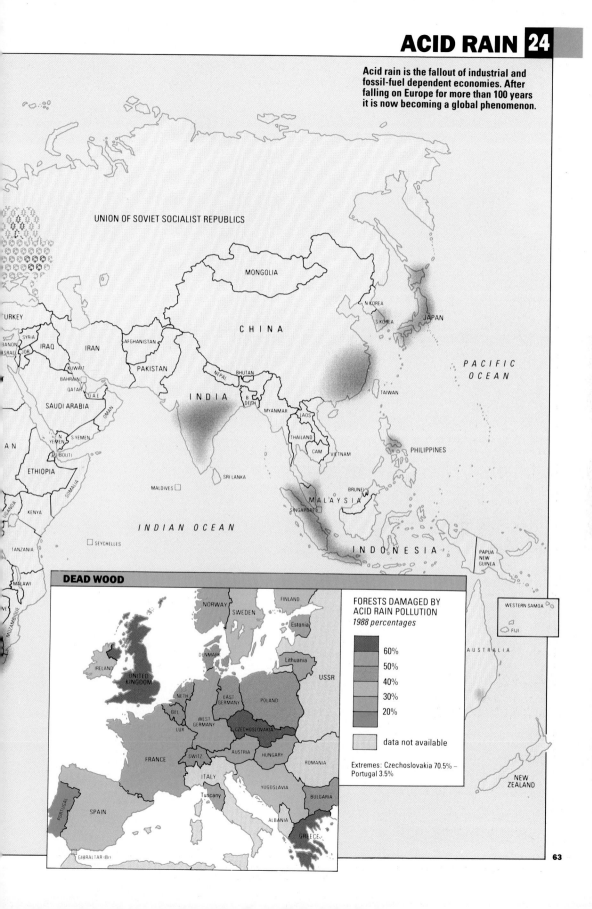

UNION OF SOVIET SOCIALIST REPUBLICS

MONGOLIA

N KOREA

S KOREA · JAPAN

CHINA

PACIFIC OCEAN

TURKEY

SYRIA
LEBANON
ISRAEL · JOR
IRAQ
IRAN
AFGHANISTAN

TAIWAN

KUWAIT
BAHRAIN
QATAR
UAE
OMAN
SAUDI ARABIA

PAKISTAN

NEPAL BHUTAN

INDIA
B DESH
MYANMAR
LAOS

THAILAND
CAM VIETNAM

PHILIPPINES

N YEMEN S YEMEN
DJIBOUTI

SRI LANKA

ETHIOPIA
SOMALIA

MALDIVES

BRUNEI

MALAYSIA
SINGAPORE

UGANDA
KENYA

INDIAN OCEAN

INDONESIA

TANZANIA

SEYCHELLES

PAPUA NEW GUINEA

MALAWI

MOZAMBIQUE

WESTERN SAMOA

FIJI

AUSTRALIA

DEAD WOOD

FINLAND

NORWAY
SWEDEN

Estonia

DENMARK

Lithuania

IRELAND

UNITED KINGDOM

USSR

NETH
BEL
LUX
EAST GERMANY
WEST GERMANY
POLAND

CZECHOSLOVAKIA

FRANCE
SWITZ
AUSTRIA
HUNGARY

ROMANIA

ITALY

Tuscany

YUGOSLAVIA

BULGARIA

PORTUGAL
SPAIN

ALBANIA

GREECE

GIBRALTAR (Br)

FORESTS DAMAGED BY
ACID RAIN POLLUTION
1988 percentages

- 60%
- 50%
- 40%
- 30%
- 20%

data not available

Extremes: Czechoslovakia 70.5% –
Portugal 3.5%

NEW ZEALAND

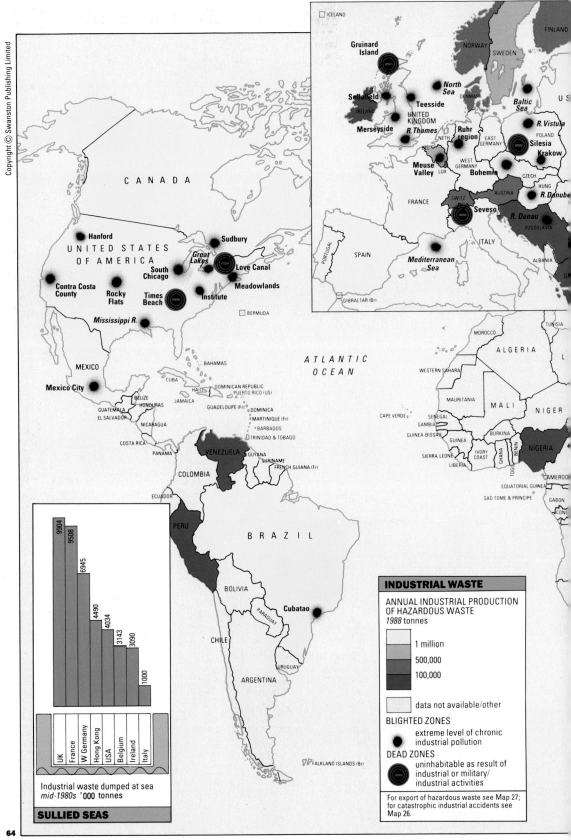

Data compiled by Joni Seager   Main sources: National Wildlife Federation *The Toxic 500 1987* (1989); UNEP *Environmental Data Report* (1989).

Chronic pollution and a worldwide trail of hazardous waste are hallmarks of the industrial age.

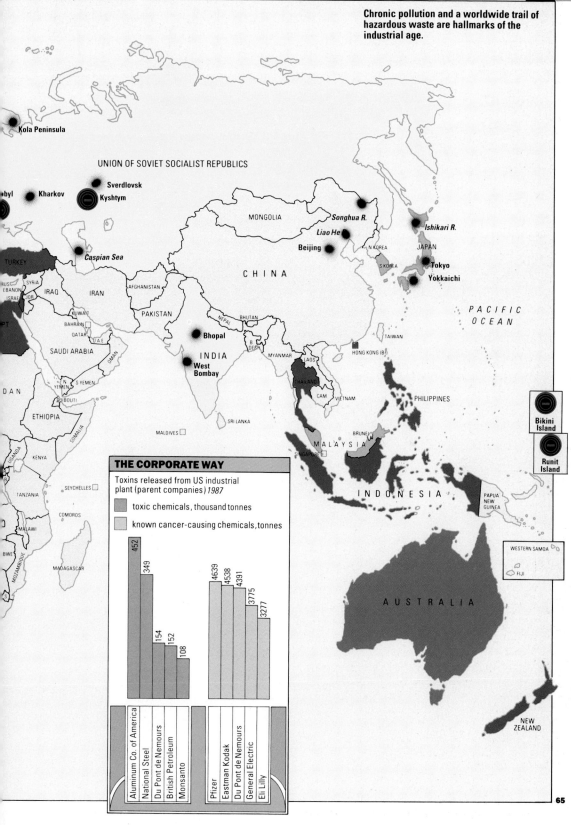

Kola Peninsula

UNION OF SOVIET SOCIALIST REPUBLICS

Sverdlovsk
byl   Kharkov   Kyshtym

MONGOLIA   Songhua R.   Ishikari R.
Liao He
Caspian Sea   Beijing   JAPAN
TURKEY   N KOREA
RUS   SYRIA   AFGHANISTAN   CHINA   S KOREA   Tokyo
EBANON   IRAQ   IRAN   Yokkaichi
ISRAEL
JOR
PT   KUWAIT   PAKISTAN   PACIFIC
BAHRAIN   NEPAL   BHUTAN   OCEAN
QATAR   UAE   Bhopal
SAUDI ARABIA   OMAN   INDIA   B   TAIWAN
DESH
West   MYANMAR   LAOS   HONG KONG (Br)
Bombay
DAN   N   S YEMEN   THAILAND
YEMEN
DJIBOUTI   CAM   VIETNAM   PHILIPPINES
ETHIOPIA   SRI LANKA   Bikini
SOMALIA   MALDIVES   Island
UGANDA
KENYA   BRUNEI   Runit
SEYCHELLES   MALAYSIA   Island
TANZANIA   SINGAPORE
COMOROS
INDONESIA   PAPUA
MALAWI   NEW
GUINEA
BWE
MADAGASCAR   WESTERN SAMOA
MOZAMBIQUE
S   FIJI

AUSTRALIA

NEW
ZEALAND

THE CORPORATE WAY

Toxins released from US industrial plant (parent companies) *1987*

| | toxic chemicals, thousand tonnes |
| | known cancer-causing chemicals, tonnes |

Aluminum Co. of America	452
National Steel	349
Du Pont de Nemours	154
British Petroleum	152
Monsanto	108
Pfizer	4639
Eastman Kodak	4538
Du Pont de Nemours	4391
General Electric	3775
Eli Lilly	3277

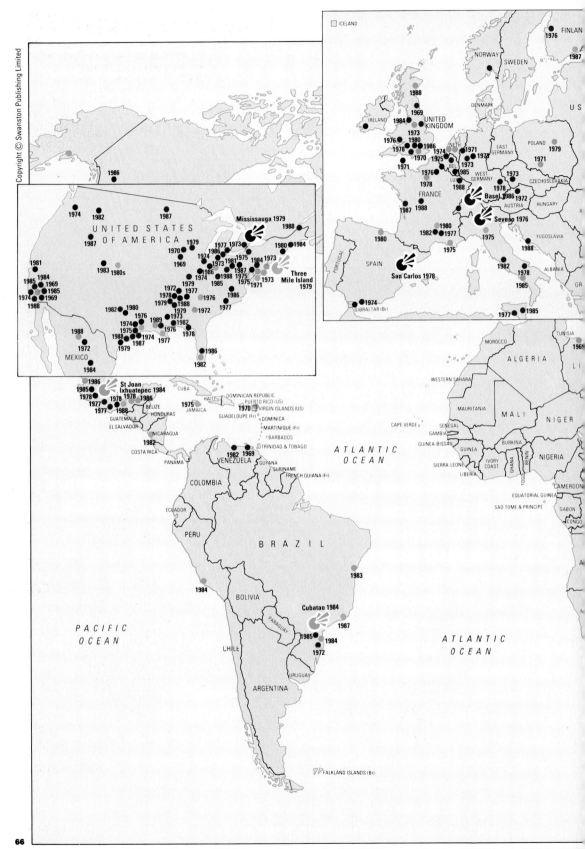

66

Data compiled by Dominic Golding   Main sources: OECD *Environmental Data Compendium* (1989); UNEP *Environmental Data Report* (1988/90); Lagadec (1982); Gittus (1988)

Accidents are usually characterized as unprecedented events. But as industrialization spreads and the products of industry become more toxic, 'accidents' are becoming a normal part of industrial culture.

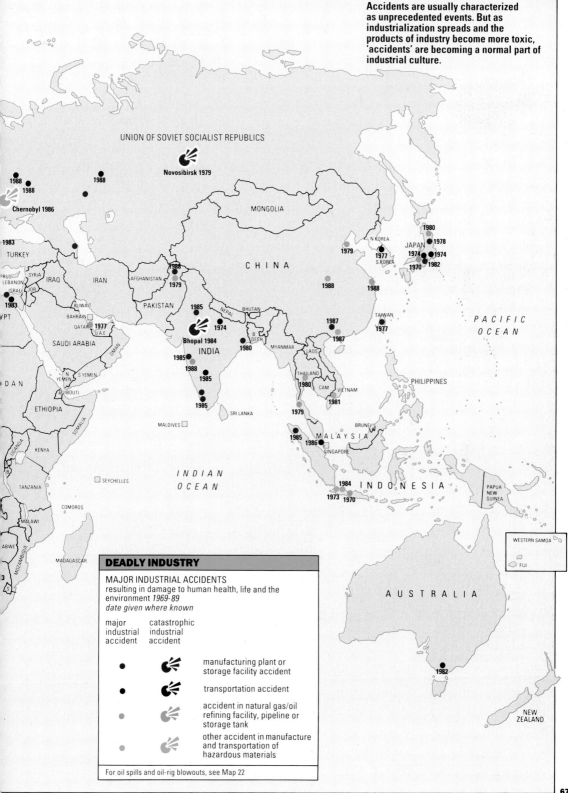

UNION OF SOVIET SOCIALIST REPUBLICS

Novosibirsk 1979

1988 1988
Chernobyl 1986

1988

MONGOLIA

1983
TURKEY

PRUS.
LEBANON SYRIA
ISRAEL JOR
1983

IRAQ IRAN AFGHANISTAN

1988
1979

C H I N A

N.KOREA
1979

1980
JAPAN 1978
1977 1974 1974
S.KOREA 1970 1982

1988 1988

KUWAIT BAHRAIN
QATAR 1977
U.A.E.
SAUDI ARABIA OMAN

PAKISTAN 1985 NEPAL BHUTAN
1974
Bhopal 1984 B.
1985 INDIA DESH
1988 1980 MYANMAR
1985

1987
1987

TAIWAN
1977

PACIFIC
OCEAN

LAOS
THAILAND
1980 CAM VIETNAM
1981

PHILIPPINES

1985
1985
1985

SRI LANKA

1979

MALDIVES

INDIAN
OCEAN

1985 MALAYSIA
1986
SINGAPORE

BRUNEI

1984 INDONESIA
1973 1970

PAPUA
NEW
GUINEA

SEYCHELLES

N
S YEMEN
YEMEN
DJIBOUTI

ETHIOPIA
SOMALIA

UGANDA
KENYA

TANZANIA

COMOROS

MALAWI

ABWE
MOZAMBIQUE

MADAGASCAR

WESTERN SAMOA

FIJI

A U S T R A L I A

1982

NEW
ZEALAND

DEADLY INDUSTRY

MAJOR INDUSTRIAL ACCIDENTS
resulting in damage to human health, life and the
environment *1969-89*
date given where known

major industrial accident	catastrophic industrial accident	
●	💥	manufacturing plant or storage facility accident
●	💥	transportation accident
●	💥	accident in natural gas/oil refining facility, pipeline or storage tank
●	💥	other accident in manufacture and transportation of hazardous materials

For oil spills and oil-rig blowouts, see Map 22

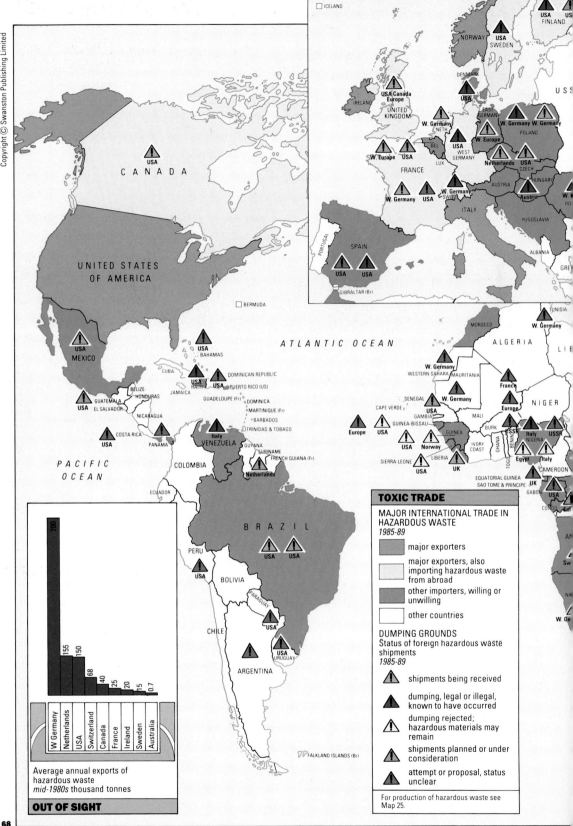

Copyright © Swanston Publishing Limited

TOXIC TRADE

MAJOR INTERNATIONAL TRADE IN HAZARDOUS WASTE
1985-89

- major exporters
- major exporters, also importing hazardous waste from abroad
- other importers, willing or unwilling
- other countries

DUMPING GROUNDS
Status of foreign hazardous waste shipments
1985-89

- shipments being received
- dumping, legal or illegal, known to have occurred
- dumping rejected; hazardous materials may remain
- shipments planned or under consideration
- attempt or proposal, status unclear

For production of hazardous waste see Map 25.

Average annual exports of hazardous waste *mid-1980s* thousand tonnes

W Germany	Netherlands	USA	Switzerland	Canada	France	Ireland	Sweden	Australia
700	155	150	68	40	25	20	15	0.7

OUT OF SIGHT

Data compiled by Marina Alberti   Main sources: Greenpeace (1989); OECD *Environmental Data Compendium* (1989).

Poor countries have become dumping grounds as industrialized countries try to ship their toxic wastes out of sight out of mind.

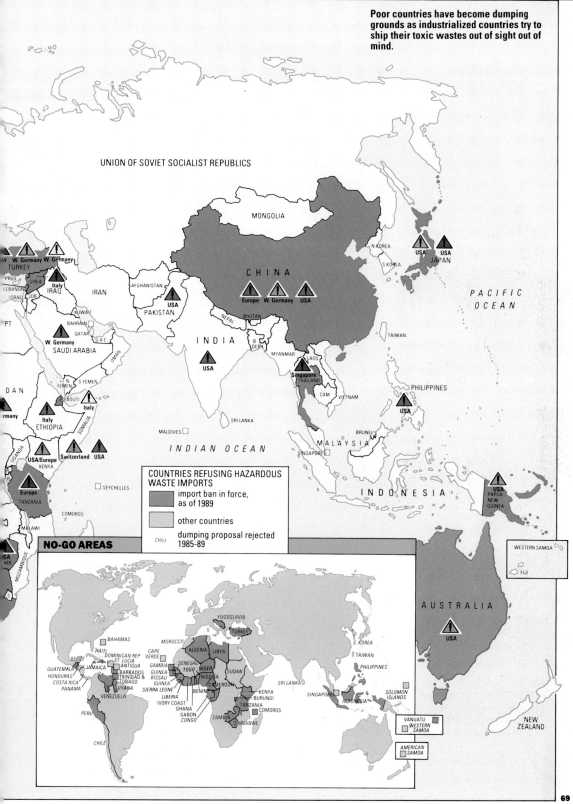

UNION OF SOVIET SOCIALIST REPUBLICS

MONGOLIA

W. Germany W. Germany
TURKEY

Italy
IRAQ

C H I N A

Europe W. Germany USA

N KOREA

USA USA
JAPAN

S KOREA

PACIFIC OCEAN

CYPRUS
LEBANON SYRIA
ISRAEL JOR

IRAN

AFGHANISTAN

USA
PAKISTAN

NEPAL BHUTAN

TAIWAN

EGYPT

BAHRAIN
QATAR U.A.E.

KUWAIT

I N D I A

B'DESH

MYANMAR

LAOS

PHILIPPINES

W. Germany
SAUDI ARABIA

OMAN

USA

Singapore
THAILAND

USA

N YEMEN S YEMEN

DJIBOUTI

Italy

SRI LANKA

CAM VIETNAM

SUDAN

Italy
ETHIOPIA

SOMALIA

MALDIVES

I N D I A N   O C E A N

BRUNEI

M A L A Y S I A

Germany

UGANDA

USA/Europe Switzerland USA
KENYA

SINGAPORE

SEYCHELLES

I N D O N E S I A

USA
PAPUA NEW GUINEA

Europe
TANZANIA

COMOROS

MALAWI

COUNTRIES REFUSING HAZARDOUS WASTE IMPORTS

import ban in force, as of 1989

other countries

dumping proposal rejected 1985-89

CHILE

WESTERN SAMOA

FIJI

SA
MB

MOZAMBIQUE

NO-GO AREAS

A U S T R A L I A

USA

BAHAMAS

YUGOSLAVIA
TURKEY

MOROCCO

HAITI
DOMINICAN REP
ST. LUCIA
ANTIGUA
BARBADOS
TRINIDAD & TOBAGO

CAPE VERDE

ALGERIA LIBYA

S. KOREA

TAIWAN

BELIZE
GUATEMALA
HONDURAS
COSTA RICA
PANAMA

JAMAICA

SENEGAL
GAMBIA
GUINEA BISSAU
GUINEA

TOGO NIGER
NIGERIA

SUDAN

PHILIPPINES

SRI LANKA

VENEZUELA

GUYANA

SIERRA LEONE
LIBERIA
IVORY COAST
GHANA

BENIN

CAMEROON

KENYA
BURUNDI

SINGAPORE

INDONESIA

SOLOMON ISLANDS

PERU

GABON
CONGO

ZAMBIA

TANZANIA
COMOROS

ZIMBABWE

VANUATU
WESTERN SAMOA

NEW ZEALAND

CHILE

AMERICAN SAMOA

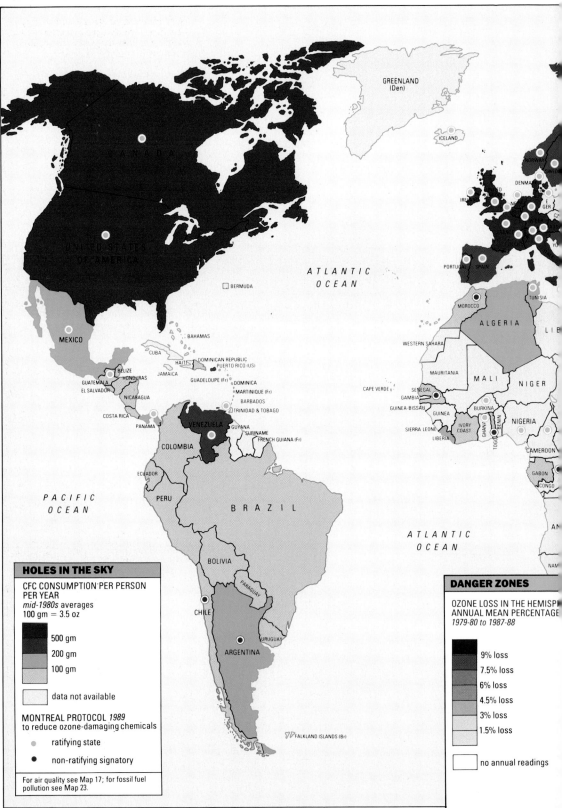

GREENLAND
(Den)

ICELAND

NORWAY

DENMARK

IRE

GER

CZ

PORTUGAL   SPAIN

ATLANTIC
OCEAN

BERMUDA

TUNISIA

MOROCCO

ALGERIA   LIB

WESTERN SAHARA

MAURITANIA   MALI   NIGER

CAPE VERDE

SENEGAL

GAMBIA

BURKINA

GUINEA-BISSAU

GUINEA   NIGERIA

SIERRA LEONE   IVORY   GHANA BENIN

COAST   TOGO

LIBERIA

CAMEROON

GABON

CONGO

MEXICO

BAHAMAS

CUBA

HAITI

DOMINICAN REPUBLIC
PUERTO RICO (US)

BELIZE

JAMAICA

GUATEMALA
HONDURAS

EL SALVADOR

GUADELOUPE (Fr) DOMINICA

NICARAGUA

MARTINIQUE (Fr)

COSTA RICA

BARBADOS

TRINIDAD & TOBAGO

PANAMA

VENEZUELA

GUYANA

SURINAME

COLOMBIA

FRENCH GUIANA (Fr)

ECUADOR

PERU   B R A Z I L

ATLANTIC
OCEAN

PACIFIC
OCEAN

AM

BOLIVIA

NAM

PARAGUAY

CHILE

URUGUAY

ARGENTINA

FALKLAND ISLANDS (Br)

HOLES IN THE SKY

CFC CONSUMPTION PER PERSON
PER YEAR
mid-1980s averages
100 gm = 3.5 oz

- 500 gm
- 200 gm
- 100 gm

☐ data not available

MONTREAL PROTOCOL *1989*
to reduce ozone-damaging chemicals

- ◉ ratifying state
- ● non-ratifying signatory

For air quality see Map 17; for fossil fuel
pollution see Map 23.

DANGER ZONES

OZONE LOSS IN THE HEMISPI
ANNUAL MEAN PERCENTAGE
1979-80 to 1987-88

- 9% loss
- 7.5% loss
- 6% loss
- 4.5% loss
- 3% loss
- 1.5% loss

☐ no annual readings

Data compiled by Nigel Dudley   Main sources: US Environmental Protection Agency (1986); NASA Ozone Trends Panel;
John Gille, US National Center for Atmospheric Research.

The chemicals that are now destroying the ozone layer do not occur naturally – they are entirely the creation of modern industry.

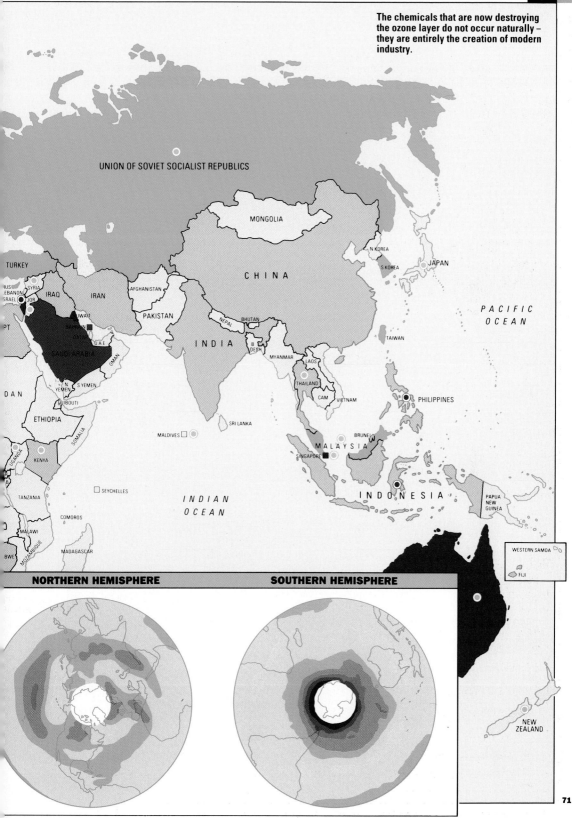

UNION OF SOVIET SOCIALIST REPUBLICS

MONGOLIA

N KOREA
S KOREA
JAPAN

CHINA

TURKEY
CRUS
LEBANON
ISRAEL
SYRIA
DOR
IRAQ
IRAN
AFGHANISTAN
PT
KWAIT
PAKISTAN
NEPAL
BHUTAN
DATAR
UAE
SAUDI ARABIA
OMAN
INDIA
B
DESH
MYANMAR
LAOS
TAIWAN

PACIFIC
OCEAN

DAN
N
YEMEN
S YEMEN
DJIBOUTI
ETHIOPIA
SOMALIA
THAILAND
CAM
VIETNAM
PHILIPPINES

SRI LANKA
MALDIVES

UGANDA
KENYA

BRUNEI
MALAYSIA
SINGAPORE

SEYCHELLES

TANZANIA

COMOROS

INDIAN
OCEAN

INDONESIA

PAPUA
NEW
GUINEA

MALAWI
MADAGASCAR
MOZAMBIQUE
BWE

WESTERN SAMOA
FIJI

NORTHERN HEMISPHERE   **SOUTHERN HEMISPHERE**

NEW
ZEALAND

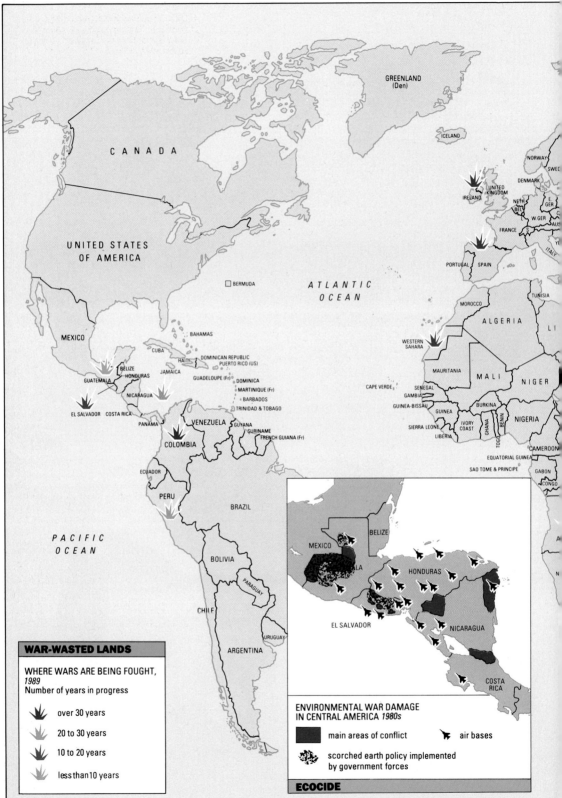

GREENLAND
(Den)

ICELAND

C A N A D A

NORWAY

SWED

DENMARK

UNITED
KINGDOM

IRELAND

NETH
BEL

E.
GER

W.GER

UNITED STATES
OF AMERICA

FRANCE

S

ITALY

PORTUGAL SPAIN

BERMUDA

A T L A N T I C
O C E A N

TUNISIA

MOROCCO

ALGERIA

LI

MEXICO

BAHAMAS

CUBA

WESTERN
SAHARA

DOMINICAN REPUBLIC
PUERTO RICO (US)

HAITI

BELIZE

JAMAICA

MAURITANIA

MALI

NIGER

HONDURAS

GUATEMALA

GUADELOUPE (Fr)

DOMINICA

MARTINIQUE (Fr)

CAPE VERDE

SENEGAL

NICARAGUA

BARBADOS

GAMBIA

GUINEA-BISSAU

BURKINA

EL SALVADOR   COSTA RICA

TRINIDAD & TOBAGO

GUINEA

BENIN

NIGERIA

PANAMA

VENEZUELA

GUYANA

SIERRA LEONE

IVORY
COAST

GHANA

TOGO

COLOMBIA

SURINAME

FRENCH GUIANA (Fr)

LIBERIA

CAMEROON

ECUADOR

EQUATORIAL GUINEA

SAO TOME & PRINCIPE

GABON

PERU

BRAZIL

CONGO

P A C I F I C
O C E A N

BOLIVIA

PARAGUAY

CHILE

MEXICO

BELIZE

LA

HONDURAS

EL SALVADOR

NICARAGUA

WAR-WASTED LANDS

WHERE WARS ARE BEING FOUGHT,
1989
Number of years in progress

URUGUAY

ARGENTINA

COSTA
RICA

over 30 years

20 to 30 years

10 to 20 years

less than 10 years

ENVIRONMENTAL WAR DAMAGE
IN CENTRAL AMERICA *1980s*

main areas of conflict

air bases

scorched earth policy implemented
by government forces

ECOCIDE

Data compiled by Dan Smith   Main sources: EPOCA (1989); SIPRI *Yearbook* (1987, 1988, 1989); Sivard (1989); Wallensteen (1988)

As the weapons of war become more powerful, so the environmental damage they cause escalates. Often the environment itself is a military target.

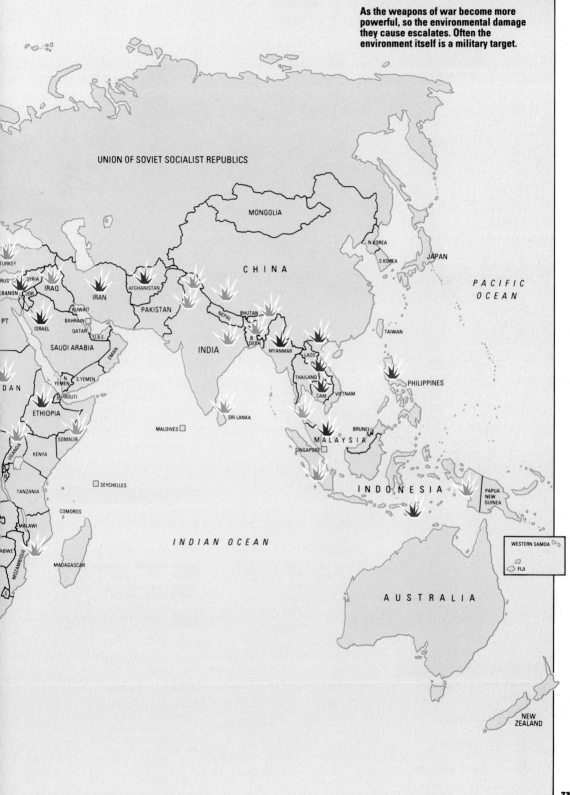

UNION OF SOVIET SOCIALIST REPUBLICS

MONGOLIA

N.KOREA

S.KOREA

JAPAN

CHINA

PACIFIC OCEAN

TURKEY

RUS

SYRIA

BANON

JOR

IRAQ

IRAN

AFGHANISTAN

KUWAIT

BAHRAIN

QATAR

UAE

OMAN

PAKISTAN

NEPAL

BHUTAN

B
DESH

MYANMAR

LAOS

TAIWAN

PT

ISRAEL

SAUDI ARABIA

INDIA

THAILAND

VIETNAM

CAM

PHILIPPINES

N
YEMEN

S.YEMEN

DJIBOUTI

SRI LANKA

MALDIVES

DAN

ETHIOPIA

SOMALIA

KENYA

BRUNEI

MALAYSIA

SINGAPORE

UGANDA

B

TANZANIA

SEYCHELLES

COMOROS

INDONESIA

PAPUA
NEW
GUINEA

MALAWI

INDIAN OCEAN

WESTERN SAMOA

FIJI

ABWE

MOZAMBIQUE

MADAGASCAR

AUSTRALIA

S

NEW
ZEALAND

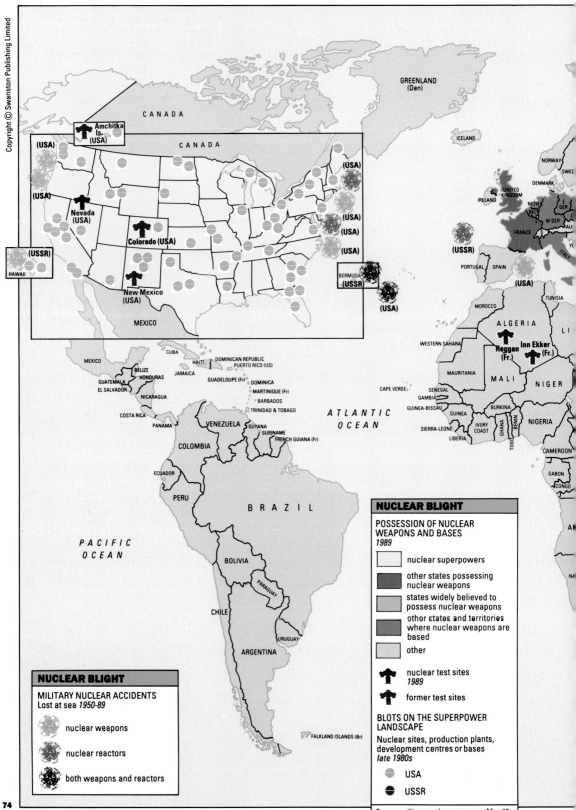

Copyright © Swanston Publishing Limited

GREENLAND (Den)

CANADA

CANADA

✚ Amchitka Is. (USA)

(USA)

(USA)

✚ Nevada (USA)

✚ Colorado (USA)

(USSR)

HAWAII

✚ New Mexico (USA)

MEXICO

(USA)

(USA)

(USA)

(USA)

(USA)

BERMUDA (USSR)

(USA)

ICELAND

NORWAY

SWED

DENMARK

IRELAND

UNITED KINGDOM

NETH

BEL

E GER

W GER

FRANCE

S

AU

(USSR)

PORTUGAL

SPAIN

ITALY

Y

(USA)

MOROCCO

TUNISIA

ALGERIA

LI

WESTERN SAHARA

Reggan (Fr.) ✚

Inn Ekker (Fr.) ✚

MAURITANIA

MALI

NIGER

MEXICO

CUBA

HAITI

DOMINICAN REPUBLIC

PUERTO RICO (US)

BELIZE

JAMAICA

GUADELOUPE (Fr)

HONDURAS

GUATEMALA

EL SALVADOR

NICARAGUA

DOMINICA

MARTINIQUE (Fr)

BARBADOS

TRINIDAD & TOBAGO

CAPE VERDE

SENEGAL

GAMBIA

GUINEA-BISSAU

GUINEA

BURKINA

SIERRA-LEONE

IVORY COAST

GHANA

BENIN

NIGERIA

LIBERIA

TOGO

CAMEROON

COSTA RICA

PANAMA

VENEZUELA

GUYANA

SURINAME

FRENCH GUIANA (Fr)

ATLANTIC OCEAN

COLOMBIA

ECUADOR

GABON

CONGO

PERU

B R A Z I L

AN

PACIFIC OCEAN

BOLIVIA

PARAGUAY

N

CHILE

URUGUAY

ARGENTINA

FALKLAND ISLANDS (Br)

NUCLEAR BLIGHT

POSSESSION OF NUCLEAR WEAPONS AND BASES
1989

☐ nuclear superpowers

■ other states possessing nuclear weapons

▨ states widely believed to possess nuclear weapons

▨ other states and territories where nuclear weapons are based

☐ other

✚ nuclear test sites *1989*

✚ former test sites

BLOTS ON THE SUPERPOWER LANDSCAPE

Nuclear sites, production plants, development centres or bases *late 1980s*

◖ USA

◗ USSR

NUCLEAR BLIGHT

MILITARY NUCLEAR ACCIDENTS
Lost at sea *1950-89*

nuclear weapons

nuclear reactors

both weapons and reactors

Data compiled by Dan Smith   Main sources: Arkin and Fieldhouse (1985); Arkin and Handler (1989); *SIPRI Yearbooks*

For non-military nuclear power see Map 19.

The nuclear activities of a few military states pose a real and present danger to the global environment.

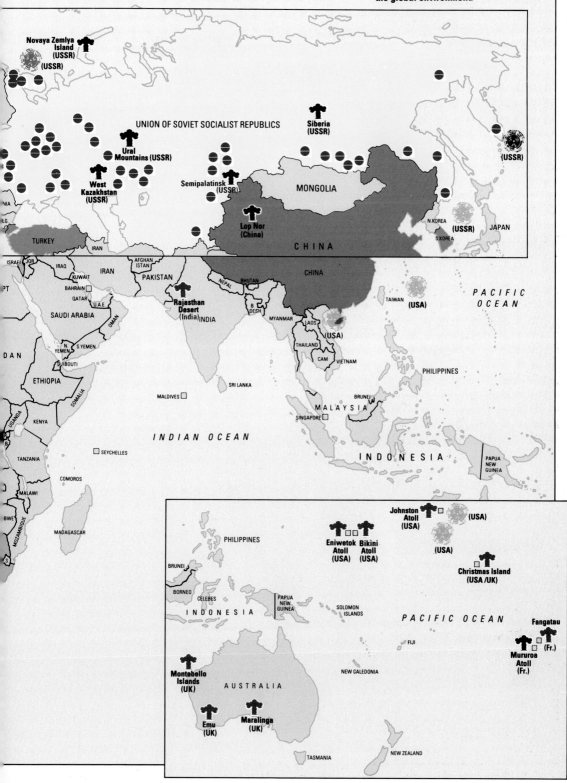

Novaya Zemlya Island (USSR) (USSR)

UNION OF SOVIET SOCIALIST REPUBLICS

Siberia (USSR)

Ural Mountains (USSR)

West Kazakhstan (USSR)

Semipalatinsk (USSR)

MONGOLIA

(USSR)

N.KOREA (USSR)

S.KOREA

JAPAN

Lop Nor (China)

TURKEY

IRAN

CHINA

ISRAEL JOR IRAQ KUWAIT IRAN AFGHANISTAN PAKISTAN NEPAL BHUTAN CHINA

BAHRAIN QATAR U.A.E.

SAUDI ARABIA OMAN

Rajasthan Desert (India) INDIA

B DESH MYANMAR

TAIWAN (USA)

PACIFIC OCEAN

N.YEMEN S.YEMEN DJIBOUTI

ETHIOPIA SOMALIA

LAOS (USA)

THAILAND CAM VIETNAM

UGANDA KENYA

SRI LANKA

PHILIPPINES

MALDIVES

BRUNEI

TANZANIA

SEYCHELLES

INDIAN OCEAN

MALAYSIA SINGAPORE

COMOROS

INDONESIA

PAPUA NEW GUINEA

MALAWI

MADAGASCAR

MOZAMBIQUE

Johnston Atoll (USA)

PHILIPPINES

Eniwetok Atoll (USA)

Bikini Atoll (USA)

(USA)

(USA)

Christmas Island (USA /UK)

BRUNEI

BORNEO CELEBES

PAPUA NEW GUINEA

INDONESIA

SOLOMON ISLANDS

PACIFIC OCEAN

Fangatau

FIJI

Mururoa Atoll (Fr.)

(Fr.)

Montebello Islands (UK)

AUSTRALIA

NEW CALEDONIA

Emu (UK)

Maralinga (UK)

TASMANIA

NEW ZEALAND

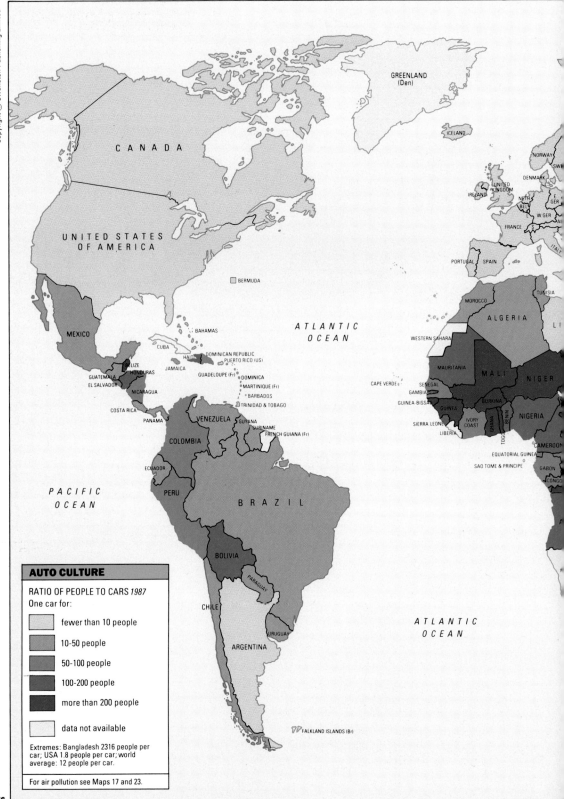

GREENLAND
(Den)

ICELAND

C A N A D A

NORWAY
DENMARK
IRELAND UNITED
KINGDOM NETH.
BEL.
W GER
FRANCE

U N I T E D   S T A T E S
O F   A M E R I C A

PORTUGAL SPAIN

ITALY

BERMUDA

TUNISIA

MOROCCO

ATLANTIC
OCEAN

ALGERIA

WESTERN SAHARA

MEXICO

BAHAMAS

CUBA

MAURITANIA M A L I NIGER

HAITI DOMINICAN REPUBLIC
PUERTO RICO (US)

BELIZE
GUATEMALA HONDURAS JAMAICA
EL SALVADOR
NICARAGUA

DOMINICA

CAPE VERDE

SENEGAL

GAMBIA

GUINEA-BISSAU

BURKINA

GUADELOUPE (Fr)

MARTINIQUE (Fr)

BARBADOS

TRINIDAD & TOBAGO

GUINEA IVORY
COAST GHANA BENIN NIGERIA

SIERRA LEONE

LIBERIA TOGO CAMEROON

COSTA RICA

PANAMA

VENEZUELA GUYANA

SURINAME

FRENCH GUIANA (Fr)

EQUATORIAL GUINEA

SAO TOME & PRINCIPE GABON

COLOMBIA

CONGO

ECUADOR

PACIFIC
OCEAN

PERU

B R A Z I L

BOLIVIA

PARAGUAY

CHILE

ATLANTIC
OCEAN

URUGUAY

ARGENTINA

AUTO CULTURE

RATIO OF PEOPLE TO CARS *1987*
One car for:

	fewer than 10 people
	10-50 people
	50-100 people
	100-200 people
	more than 200 people
	data not available

Extremes: Bangladesh 2316 people per
car; USA 1.8 people per car; world
average: 12 people per car.

For air pollution see Maps 17 and 23.

FALKLAND ISLANDS (Br)

Data compiled by Joni Seager   Main sources: *World Motor Vehicle Data* (1989).

The number of cars in the world has doubled in the past 20 years.

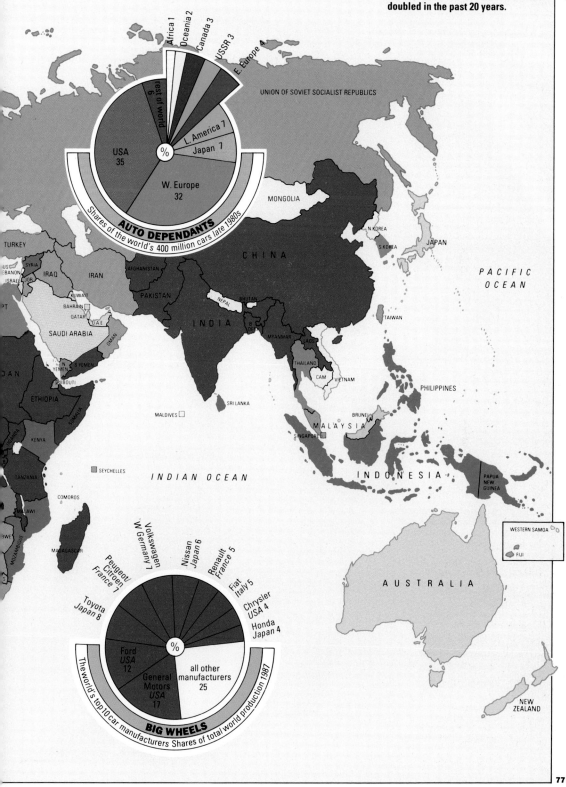

Africa 1
Oceania 2
Canada 3
USSR 3
E. Europe 4

UNION OF SOVIET SOCIALIST REPUBLICS

rest of world 6

L. America 7

Japan 7

USA 35

W. Europe 32

%

AUTO DEPENDANTS
Shares of the world's 400 million cars late 1980s

TURKEY

MONGOLIA

US
LEBANON
ISRAEL
JOR
SYRIA
IRAQ
IRAN

PT

KUWAIT
BAHRAIN
QATAR
U.A.E.
OMAN
SAUDI ARABIA

AFGHANISTAN

PAKISTAN

CHINA

N KOREA
S KOREA

JAPAN

PACIFIC
OCEAN

NEPAL
BHUTAN

INDIA

TAIWAN

DAN

N YEMEN
S YEMEN
DJIBOUTI

ETHIOPIA

MALDIVES

SRI LANKA

B DESH
MYANMAR
LAOS
THAILAND
CAM
VIETNAM

PHILIPPINES

BRUNEI
MALAYSIA

SINGAPORE

INDONESIA

PAPUA
NEW
GUINEA

KENYA

SEYCHELLES

INDIAN OCEAN

TANZANIA

COMOROS

MALAWI

BWE

MOZAMBIQUE

MADAGASCAR

WESTERN SAMOA

FIJI

AUSTRALIA

Volkswagen
W Germany 7

Nissan
Japan 6

Renault
France 5

Peugeot
Citroen
France 7

Fiat
Italy 5

Chrysler
USA 4

Toyota
Japan 8

Honda
Japan 4

Ford
USA
12

%

all other
manufacturers
25

General
Motors
USA
17

BIG WHEELS
The world's top 10 car manufacturers Shares of total world production 1987

NEW
ZEALAND

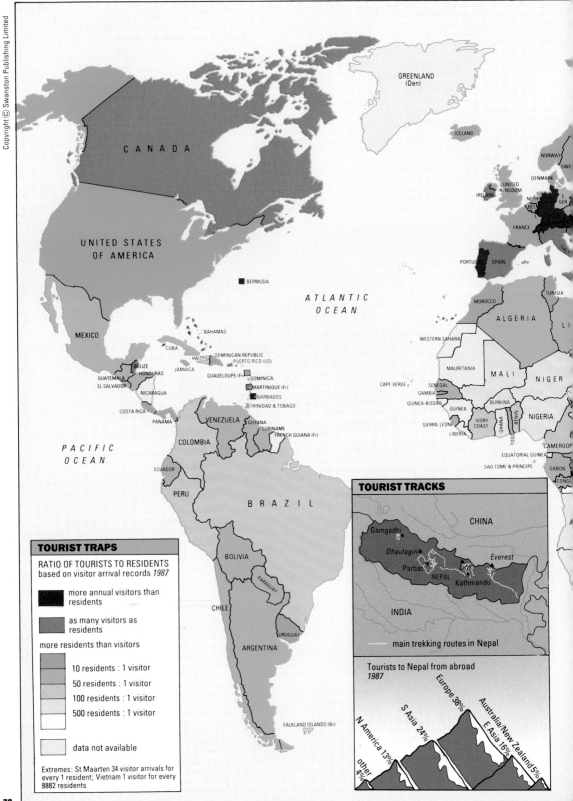

GREENLAND
(Den)

ICELAND

NORWAY

SWE

C A N A D A

DENMARK

UNITED
KINGDOM

IRELAND

NETH
BEL
L

E
GER

FRANCE

UNITED STATES
OF AMERICA

BERMUDA

PORTUGAL SPAIN

ITALY

A T L A N T I C
O C E A N

TUNISIA

MOROCCO

ALGERIA

LI

MEXICO

BAHAMAS

WESTERN SAHARA

CUBA

HAITI

DOMINICAN REPUBLIC
PUERTO RICO (US)

BELIZE

HONDURAS

JAMAICA

GUADELOUPE (Fr)

GUATEMALA

DOMINICA

EL SALVADOR

MARTINIQUE (Fr)

NICARAGUA

BARBADOS

TRINIDAD & TOBAGO

COSTA RICA

PANAMA

MAURITANIA

M A L I

NIGER

CAPE VERDE

SENEGAL

GAMBIA

GUINEA-BISSAU

GUINEA

BURKINA

NIGERIA

SIERRA LEONE

IVORY
COAST

GHANA

TOGO

BENIN

LIBERIA

VENEZUELA

GUYANA

SURINAME

FRENCH GUIANA (Fr)

COLOMBIA

CAMEROON

EQUATORIAL GUINEA

SAO TOME & PRINCIPE

GABON

CONGO

P A C I F I C
O C E A N

ECUADOR

PERU

B R A Z I L

TOURIST TRACKS

CHINA

Gamgadhi

Dhaulagiri▲

Parbat

▲

Everest

BOLIVIA

NEPAL

Kathmandu

PARAGUAY

INDIA

CHILE

— main trekking routes in Nepal

URUGUAY

ARGENTINA

TOURIST TRAPS

RATIO OF TOURISTS TO RESIDENTS
based on visitor arrival records *1987*

more annual visitors than
residents

as many visitors as
residents

more residents than visitors

10 residents : 1 visitor

50 residents : 1 visitor

100 residents : 1 visitor

500 residents : 1 visitor

data not available

Extremes: St Maarten 34 visitor arrivals for
every 1 resident; Vietnam 1 visitor for every
8882 residents

FALKLAND ISLANDS (Br)

Tourists to Nepal from abroad
1987

Europe 38%

Australia/New
Zealand 5%

S Asia 24%

E Asia 16%

N America 13%

other
4%

Data compiled by Nigel Dudley and Mark Davis   Main sources: Commission of the European Communities *Quality of Bathing Water* (1987); World Tourism Organisation (198

The needs of tourists, especially rich tourists in poor countries, can deplete local resources and leave a global trail of litter and waste.

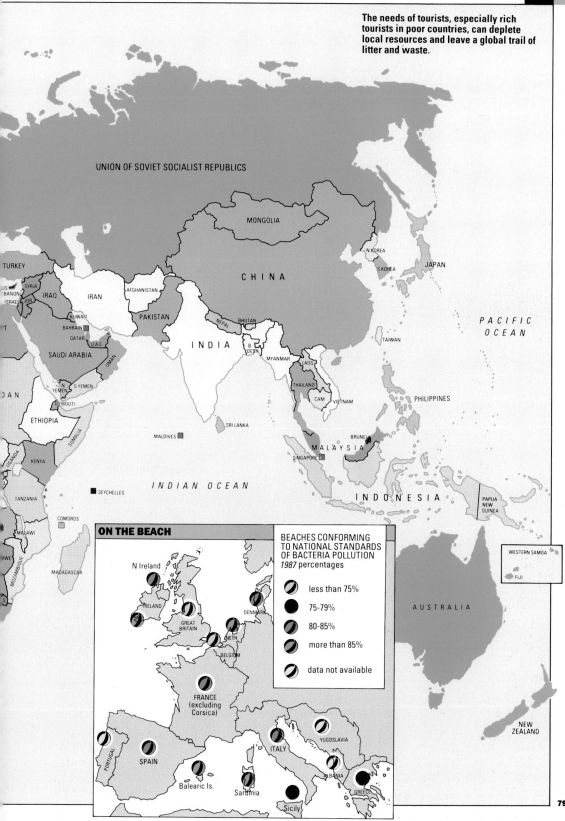

UNION OF SOVIET SOCIALIST REPUBLICS

MONGOLIA

N KOREA

S KOREA

JAPAN

TURKEY

SYRIA

LEBANON

IRAQ

IRAN

AFGHANISTAN

CHINA

PACIFIC OCEAN

ISRAEL   JOR

KUWAIT

BAHRAIN

QATAR

U.A.E

PAKISTAN

NEPAL

BHUTAN

INDIA

B DESH

MYANMAR

LAOS

TAIWAN

SAUDI ARABIA

OMAN

THAILAND

CAM

VIETNAM

PHILIPPINES

N YEMEN

S YEMEN

DJIBOUTI

ETHIOPIA

SRI LANKA

MALDIVES

BRUNEI

MALAYSIA

SINGAPORE

UGANDA

KENYA

SOMALIA

SEYCHELLES

INDIAN OCEAN

INDONESIA

PAPUA NEW GUINEA

TANZANIA

COMOROS

MALAWI

MADAGASCAR

MOZAMBIQUE

WESTERN SAMOA

FIJI

AUSTRALIA

NEW ZEALAND

ON THE BEACH

BEACHES CONFORMING TO NATIONAL STANDARDS OF BACTERIA POLLUTION
1987 percentages

- less than 75%
- 75-79%
- 80-85%
- more than 85%
- data not available

N Ireland

IRELAND

GREAT BRITAIN

DENMARK

NETH

BELGIUM

FRANCE (excluding Corsica)

YUGOSLAVIA

ITALY

PORTUGAL

SPAIN

Balearic Is.

Sardinia

ALBANIA

GREECE

Sicily

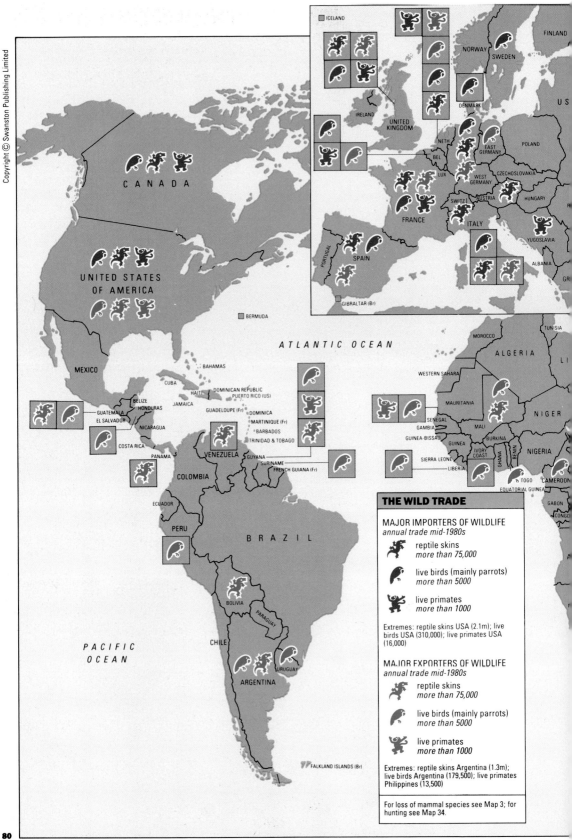

THE WILD TRADE

MAJOR IMPORTERS OF WILDLIFE
annual trade mid-1980s

reptile skins
more than 75,000

live birds (mainly parrots)
more than 5000

live primates
more than 1000

Extremes: reptile skins USA (2.1m); live
birds USA (310,000); live primates USA
(16,000)

MAJOR EXPORTERS OF WILDLIFE
annual trade mid-1980s

reptile skins
more than 75,000

live birds (mainly parrots)
more than 5000

live primates
more than 1000

Extremes: reptile skins Argentina (1.3m);
live birds Argentina (179,500); live primates
Philippines (13,500)

For loss of mammal species see Map 3; for
hunting see Map 34.

Data compiled by Andy Crump   Main sources: World Resources Institute (1988); IUCN Monitoring Centre

The trade in wildlife and wildlife products is a $5 billion industry that threatens whole species and ecosystems.

THE IVORY TRADERS

USA
SUDAN
CAR
TANZANIA
BOTSWANA
ZIMBABWE
SOUTH AFRICA
U.A.E
INDIA
JAPAN
HONG KONG (Br)
SINGAPORE

top 12 ivory traders

THE FUR TRADERS

CANADA
USA
UNITED KINGDOM
FRANCE
DENMARK
W.GER
ITALY
CHINA
HONG KONG
BOLIVIA
BOTSWANA
SOUTH AFRICA

top 12 wildcat-fur traders

PACIFIC OCEAN

SSR
TURKEY
RUSS
SYRIA
EBANON
ISRAEL
IRAQ
IRAN
AFGHANISTAN
CHINA
N KOREA
S.KOREA
JAPAN
KUWAIT
BAHRAIN
QATAR
U.A.E
SAUDI ARABIA
OMAN
PAKISTAN
NEPAL
BHUTAN
INDIA
B DESH
MYANMAR
LAOS
THAILAND
CAM
VIETNAM
TAIWAN
N YEMEN
S YEMEN
DJIBOUTI
DAN
ETHIOPIA
SOMALIA
SRI LANKA
MALDIVES
PHILIPPINES
UGANDA
KENYA
BRUNEI
MALAYSIA
SINGAPORE
SEYCHELLES
TANZANIA
INDIAN OCEAN
INDONESIA
PAPUA NEW GUINEA
COMOROS
MALAWI
MADAGASCAR
BWE
MOZAMBIQUE
WESTERN SAMOA
FIJI
AUSTRALIA
NEW ZEALAND

Copyright © Swanston Publishing Limited

HUNTING

WILDLIFE ENDANGERED BY HUNTING *mid-1980s*

Endangered species

mammals	
reptiles	
birds	

Species successfully recovering

mammals	
birds	

For loss of mammal species see Map 3; for trade in wildlife see Map 33.

Data compiled by Andy Crump   Main sources: Burton & Pearson ; Fitter (1986); Inskipp & Wells (1979); World Resources Institute (1987, 1988); OECD *Environmental Data Compendium* (1989); IUCN Monitoring Centre (1988).

Hunting for sport or profit is one of the greatest threats to wildlife and wilderness ecosystems.

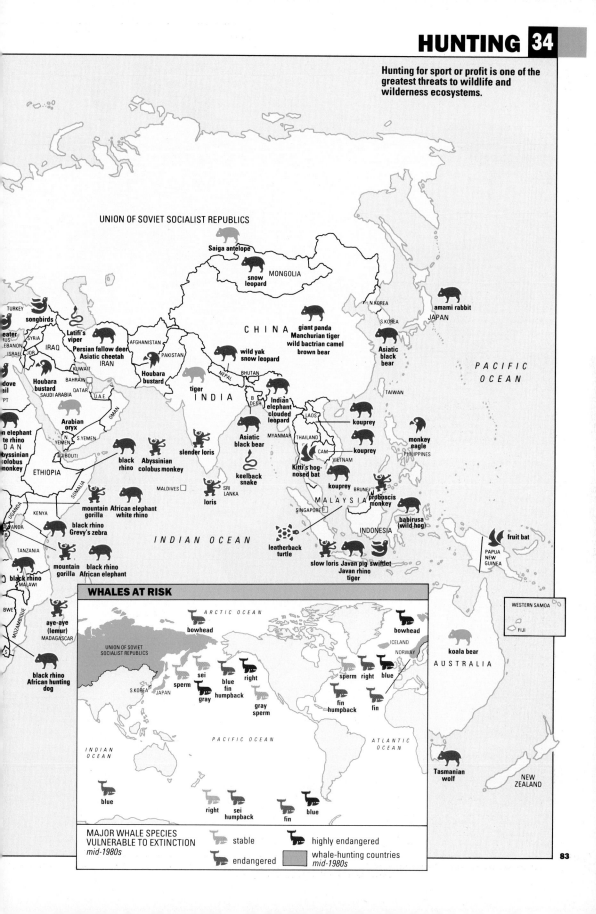

UNION OF SOVIET SOCIALIST REPUBLICS

Saiga antelope

snow leopard

MONGOLIA

N KOREA

S KOREA

amami rabbit
JAPAN

TURKEY
songbirds

eater
ail
RUS
SYRIA
LEBANON
ISRAEL
JOR

Latifi's viper

Persian fallow deer
Asiatic cheetah
IRAN

AFGHANISTAN

PAKISTAN

giant panda
Manchurian tiger
wild bactrian camel
brown bear

C H I N A

wild yak
snow leopard

Asiatic
black
bear

PACIFIC
OCEAN

IRAQ

KUWAIT

dove
ail
PT

Houbara
bustard
SAUDI ARABIA

BAHRAIN
QATAR
U.A.E.

N
YEMEN

Houbara
bustard

NEPAL

BHUTAN

tiger
I N D I A

B
DESH

Indian
elephant
clouded
leopard

LAOS

TAIWAN

kod

kouprey

MYANMAR

THAILAND

CAM

monkey
eagle
PHILIPPINES

Arabian
oryx

OMAN

n elephant
te rhino
DAN
byssinian
:olobus
monkey

ETHIOPIA

SOMALIA

DJIBOUTI

black
rhino

Abyssinian
colobus monkey

slender loris

Asiatic
black
bear

VIETNAM

kouprey

Kitti's hog-
nosed bat

keelback
snake

MALDIVES

SRI
LANKA

loris

kouprey

BRUNEI

SINGAPORE

M A L A Y S I A

proboscis
monkey

UGANDA

KENYA

mountain
gorilla

African elephant
white rhino

I N D I A N   O C E A N

babirusa
(wild hog)

INDONESIA

fruit bat

ANDA

black rhino
Grevy's zebra

leatherback
turtle

PAPUA
NEW
GUINEA

TANZANIA

mountain
gorilla

black rhino
African elephant

slow loris Javan pig swiftlet
Javan rhino
tiger

WESTERN SAMOA

FIJI

black rhino
MALAWI

BWE

aye-aye
(lemur)
MADAGASCAR

MOZAMBIQUE

black rhino
African hunting
dog

WHALES AT RISK

ARCTIC OCEAN

bowhead

bowhead

UNION OF SOVIET
SOCIALIST REPUBLICS

ICELAND

NORWAY

koala bear

AUSTRALIA

sei

right

sperm right blue

S.KOREA JAPAN

sperm

gray

blue
fin
humpback

gray
sperm

fin
humpback

fin

PACIFIC OCEAN

ATLANTIC
OCEAN

INDIAN
OCEAN

Tasmanian
wolf

NEW
ZEALAND

blue

right sei
humpback

fin

blue

MAJOR WHALE SPECIES
VULNERABLE TO EXTINCTION
mid-1980s

stable

highly endangered

endangered

whale-hunting countries
mid-1980s

CANADA

UNITED STATES OF AMERICA

Asia 79%
Latin America 19%
Africa 2%

MEXICO

BERMUDA

BAHAMAS

CUBA
HAITI
DOMINICAN REPUBLIC
PUERTO RICO (US)
JAMAICA

BELIZE
GUATEMALA
HONDURAS
EL SALVADOR
NICARAGUA
COSTA RICA
PANAMA

GUADELOUPE (Fr)
DOMINICA
MARTINIQUE (Fr)
BARBADOS
TRINIDAD & TOBAGO

VENEZUELA
COLOMBIA
ECUADOR
PERU

GUYANA
SURINAME
FRENCH GUIANA (Fr)

B R A Z I L

BOLIVIA

PARAGUAY

PACIFIC OCEAN

CHILE

URUGUAY

ARGENTINA

FALKLAND ISLANDS (Br)

ATLANTIC OCEAN

ICELAND

NORWAY

FINLAND

SWEDEN

DENMARK

Asia 71%
Africa 27%
Latin America 2%

Asia 66%
Africa 18%
Latin America 16%

Asia 72%
Africa 21%
Latin America 7%

UNITED KINGDOM

Asia 54%
Africa 45%
Latin America <1%

NETH
BEL
LUX

Latin America 3%
Asia 46%
Africa 51%

E GER
POLAND

CZECHOSLOVAKIA

WEST GERMANY

Africa 82%
Asia 18%
Latin America <1%

FRANCE

AUSTRIA
HUNGARY

Africa 99%
Asia <1%
Latin America <1%

SWITZ

YUGOSLAVIA

SPAIN

Africa 75%
Asia 24%
Latin America 1%

PORTUGAL

Africa 92%
Asia 5%
Latin America 3%

ITALY
ALBANIA

SPAIN

GIBRALTAR (Br)

MOROCCO

TUNISIA

ALGERIA

LI

WESTERN SAHARA

MAURITANIA

MALI

NIGER

CAPE VERDE

SENEGAL
GAMBIA
GUINEA-BISSAU
GUINEA
SIERRA LEONE
LIBERIA

BURKINA

IVORY COAST
GHANA
TOGO
BENIN

NIGERIA

CAMEROON

EQUATORIAL GUINEA
SAO TOME & PRINCIPE

GABON
CONGO

Data compiled by Friends of the Earth/Jan McHarry   Main sources: Nectoux & Dudley (1987); Myers & Houghton (1989).

Every year tropical timber worth more than US$8 billion is shipped from poor countries to rich, destroying 12 million acres of tropical forest.

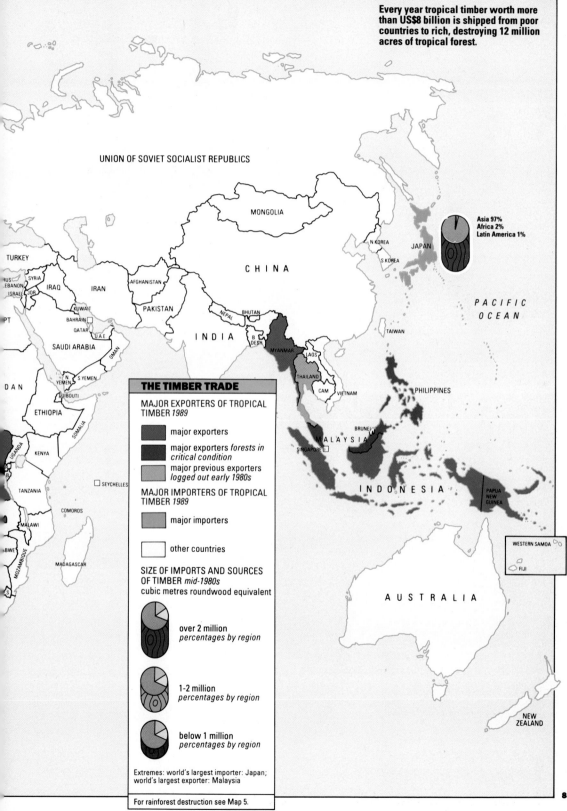

UNION OF SOVIET SOCIALIST REPUBLICS

MONGOLIA

CHINA

N KOREA
JAPAN
S KOREA

Asia 97%
Africa 2%
Latin America 1%

PACIFIC OCEAN

TURKEY

RUS
SYRIA
EBANON
ISRAEL JOR
IRAQ
IRAN
AFGHANISTAN

KUWAIT
BAHRAIN
QATAR
UAE
SAUDI ARABIA
OMAN

PAKISTAN

NEPAL
BHUTAN
B
DESH
MYANMAR

TAIWAN

INDIA

LAOS
THAILAND
CAM
VIETNAM

PHILIPPINES

PT

DAN

N
YEMEN
S YEMEN
DJIBOUTI
ETHIOPIA

SOMALIA

KENYA

BRUNEI
MALAYSIA
SINGAPORE

SEYCHELLES

UGANDA

TANZANIA

COMOROS

MALAWI

INDONESIA

PAPUA
NEW
GUINEA

BWE

MADAGASCAR

MOZAMBIQUE

WESTERN SAMOA

FIJI

AUSTRALIA

NEW
ZEALAND

THE TIMBER TRADE

MAJOR EXPORTERS OF TROPICAL TIMBER *1989*

- major exporters
- major exporters *forests in critical condition*
- major previous exporters *logged out early 1980s*

MAJOR IMPORTERS OF TROPICAL TIMBER *1989*

- major importers

- other countries

SIZE OF IMPORTS AND SOURCES OF TIMBER *mid-1980s*
cubic metres roundwood equivalent

over 2 million
percentages by region

1-2 million
percentages by region

below 1 million
percentages by region

Extremes: world's largest importer: Japan; world's largest exporter: Malaysia

For rainforest destruction see Map 5.

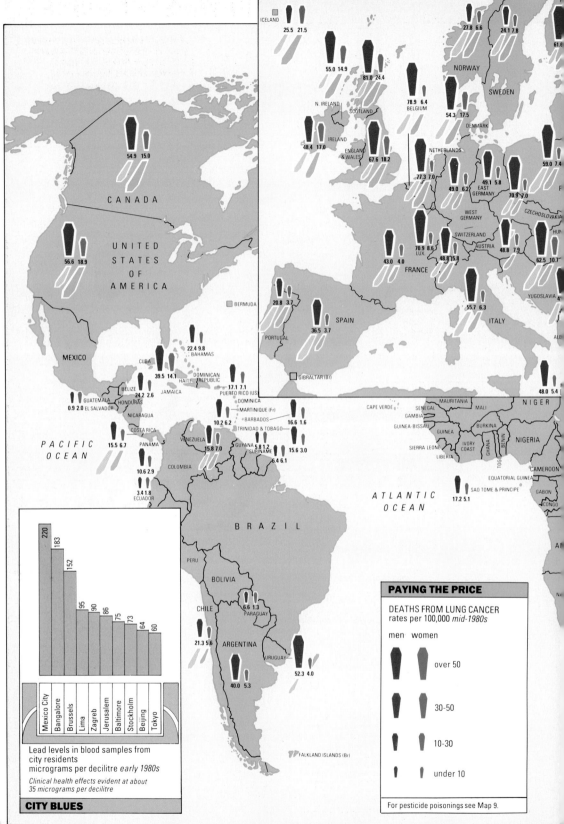

PAYING THE PRICE

DEATHS FROM LUNG CANCER
rates per 100,000 mid-1980s

men women

over 50

30-50

10-30

under 10

For pesticide poisonings see Map 9.

Lead levels in blood samples from
city residents
micrograms per decilitre early 1980s

Clinical health effects evident at about
35 micrograms per decilitre

CITY BLUES

Data compiled by Jean MacRae   Main sources: WHO, *World Health Statistics* (1988); Vater (1982).

The links between specific illnesses and environmental factors are difficult to prove. Nor can they be discounted. Smoking continues to be the most identifiable cause of lung cancer.

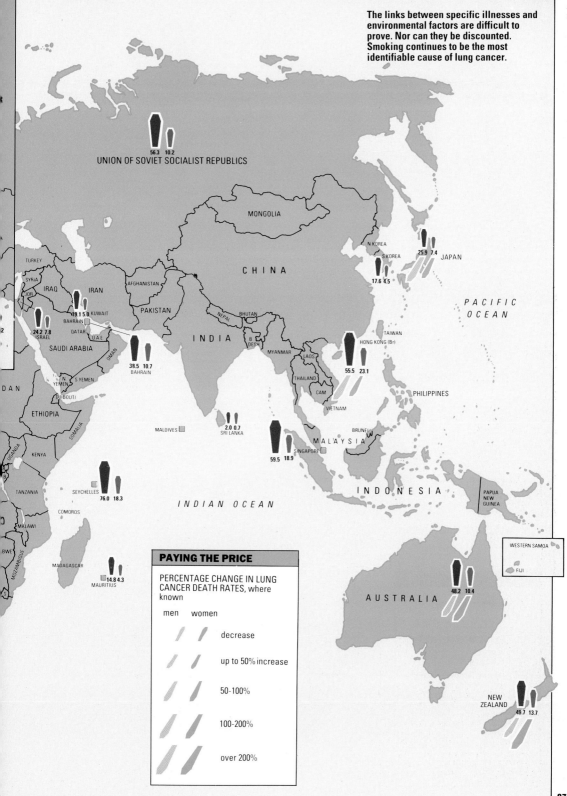

UNION OF SOVIET SOCIALIST REPUBLICS
56.3 10.2

MONGOLIA

N KOREA
S KOREA
17.6 4.5

JAPAN
25.9 7.4

TURKEY
SYRIA
JOR
IRAQ
IRAN
AFGHANISTAN
CHINA

PACIFIC
OCEAN

19.1 5.0 KUWAIT
BAHRAIN
QATAR
U.A.E
OMAN

PAKISTAN
NEPAL
BHUTAN
B DESH
INDIA
MYANMAR
LAOS
THAILAND
CAM
VIETNAM

TAIWAN
HONG KONG (Br)
55.5 23.1

24.2 7.8
ISRAEL
SAUDI ARABIA

PHILIPPINES

38.5 10.7
BAHRAIN

DAN
N YEMEN S YEMEN
DJIBOUTI
ETHIOPIA
SOMALIA

MALDIVES
2.0 0.7
SRI LANKA

BRUNEI

MALAYSIA

UGANDA
KENYA

59.5 18.9
SINGAPORE

INDONESIA

PAPUA
NEW
GUINEA

TANZANIA
SEYCHELLES
76.0 18.3

INDIAN OCEAN

COMOROS

MALAWI

BWE
MADAGASCAR
14.8 4.3
MAURITIUS

WESTERN SAMOA

FIJI

MOZAMBIQUE

48.2 10.4

AUSTRALIA

PAYING THE PRICE

PERCENTAGE CHANGE IN LUNG CANCER DEATH RATES, where known

men	women	
		decrease
		up to 50% increase
		50-100%
		100-200%
		over 200%

NEW
ZEALAND
49.7 13.7

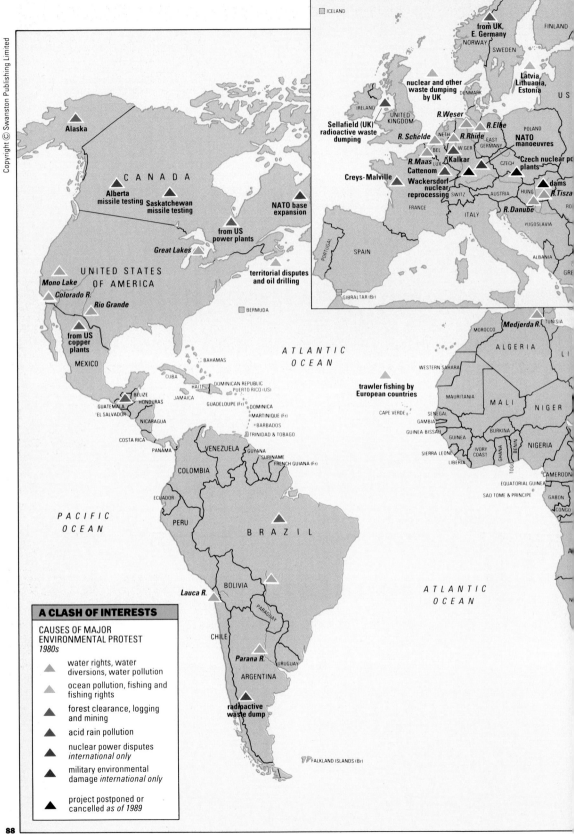

A CLASH OF INTERESTS

CAUSES OF MAJOR
ENVIRONMENTAL PROTEST
1980s

- water rights, water diversions, water pollution
- ocean pollution, fishing and fishing rights
- forest clearance, logging and mining
- acid rain pollution
- nuclear power disputes *international only*
- military environmental damage *international only*
- project postponed or cancelled *as of 1989*

Map labels:

Alaska

C A N A D A

Alberta
missile testing

Saskatchewan
missile testing

NATO base
expansion

from US
power plants

Great Lakes

UNITED STATES
OF AMERICA

Mono Lake

Colorado R.

Rio Grande

from US
copper
plants

MEXICO

territorial disputes
and oil drilling

BERMUDA

A T L A N T I C
O C E A N

BAHAMAS

CUBA

HAITI
DOMINICAN REPUBLIC
PUERTO RICO (US)

JAMAICA

BELIZE
HONDURAS
GUATEMALA
EL SALVADOR
NICARAGUA
COSTA RICA
PANAMA

GUADELOUPE (Fr)
DOMINICA
MARTINIQUE (Fr)
BARBADOS
TRINIDAD & TOBAGO

VENEZUELA
COLOMBIA
ECUADOR
GUYANA
SURINAME
FRENCH GUIANA (Fr)

PERU

B R A Z I L

trawler fishing by
European countries

CAPE VERDE

P A C I F I C
O C E A N

BOLIVIA

Lauca R.

PARAGUAY

CHILE

Parana R.

URUGUAY

ARGENTINA

radioactive
waste dump

FALKLAND ISLANDS (Br)

A T L A N T I C
O C E A N

Inset map (Europe):

ICELAND

FINLAND

from UK,
E. Germany

NORWAY

SWEDEN

nuclear and other
waste dumping
by UK

Latvia,
Lithuania,
Estonia

DENMARK

US

IRELAND

UNITED
KINGDOM

Sellafield (UK)
radioactive waste
dumping

R.Weser

R.Elhe

R.Schelde

R.Rhine

POLAND

NATO
manoeuvres

NETH
BEL
W.GER

EAST
GERMANY

CZECH

Czech nuclear po
plants

R.Maas
LUX
Kalkar

Cattenom

Creys-Malville

Wackersdorf
nuclear
reprocessing

SWITZ

AUSTRIA

HUNG

dams
R.Tisza

RO

FRANCE

ITALY

R.Danube

YUGOSLAVIA

PORTUGAL

SPAIN

ALBANIA

GRE

GIBRALTAR (Br)

Medjerda R.

TUNISIA

MOROCCO

ALGERIA

LI

WESTERN SAHARA

MAURITANIA

MALI

NIGER

SENEGAL
GAMBIA
GUINEA-BISSAU
GUINEA
SIERRA LEONE
LIBERIA
IVORY
COAST
BURKINA
GHANA
TOGO
BENIN
NIGERIA
CAMEROON

EQUATORIAL GUINEA
SAO TOME & PRINCIPE
GABON
CONGO

Data compiled by Michael Renner   Main sources: press reports.

Environmental issues are increasingly in the foreground of relations between citizens and their governments, and between states.

Siberian Rivers

UNION OF SOVIET SOCIALIST REPUBLICS

MONGOLIA

Aral Sea

driftnet fishing by Japan, S. Korea, Taiwan

N.KOREA

S.KOREA

JAPAN

TURKEY

RUSS SYRIA
LEBANON
ISRAEL JOR
IRAQ KUWAIT
BAHRAIN
QATAR U.A.E
see inset

CHINA

Three Gorges dam

PACIFIC OCEAN

R.Tigris
R.Euphrates IRAN

AFGHANISTAN

PAKISTAN

Brahmaputra R. Nu Jiang R.

Daya Bay nuclear power plant

R.Ganges NEPAL BHUTAN

INDIA

TAIWAN

HONG KONG (Br)

PT

le

SAUDI ARABIA

OMAN

Narmada Valley irrigation & hydro projects

Uttar Pradesh

B. DESH

MYANMAR LAOS

Salween R.

THAILAND

N. Luzon

Manila nuclear power plant

DAN

N. YEMEN
S. YEMEN

DJIBOUTI

Karnataka

VIETNAM

CAM

Mekong R.

PHILIPPINES

driftnet fishing by Japan, S. Korea, Taiwan

ETHIOPIA

SRI LANKA

MALDIVES

BRUNEI

UGANDA

KENYA

MALAYSIA

SINGAPORE

SEYCHELLES

INDIAN OCEAN

INDONESIA

PAPUA NEW GUINEA

TANZANIA

COMOROS

MALAWI

AUSTRALIA

WESTERN SAMOA

nuclear testing

FIJI

MADAGASCAR

ABWE

MOZAMBIQUE

TURKEY

CYPRUS

SYRIA

Mururoa Atoll nuclear testing

LEBANON

Yarmuk R.

IRAQ

West Bank

R.Jordan

ISRAEL

JORDAN

SAUDI ARABIA

Franklin dam

Tasmania pulp mill

NEW ZEALAND

	1 Population 1989 millions	2 Population growth rate 1985-90 percentages	3 Urban population 1985 percentages of total population	4 Land area protected 1988 percentages of total land	5 Carbon diox emission 1987 million ton
Afghanistan	20.2	4.83	18.5	00.2	3.99
Albania	3.3	2.10	34.0	2.0	9.45
Algeria	24.7	3.21	42.6	0.2	3.66
American Samoa (US)			48.2		
Angola	9.7	2.67	24.5	1.2	4.39
Antigua & Barbuda			30.8		0.36
Argentina	32.4	1.46	84.6	1.7	109.24
Australia	16.5	1.25	85.5	4.7	234.25
Austria	7.5	0.01	56.1	5.1	52.05
Bahamas			57.5		1.10
Bahrain	0.5	3.72	81.7	0.0	16.13
Bangladesh	112.3	2.61	11.9	0.7	12.46
Barbados	0.3	0.60	42.2	05.8	0.73
Belgium	9.9	0.09	96.3	0.4	94.94
Belize			50.0		0.18
Benin	4.6	3.12	35.2	7.6	0.36
Bermuda					0.73
Bhutan	1.5	2.03	4.5	18.6	0.0
Bolivia	7.1	2.76	47.8	4.5	4.03
Botswana	1.3	3.70	19.2	17.0	1.46
Brazil	147.4	2.07	72.7	1.4	184.40
Brunei			57.7		2.93
Bulgaria	9.2	0.38	66.5	1.0	120.61
Burkina	7.7	2.65	7.9	2.5	0.36
Burundi	5.3	2.84	5.6	0.0	0.0
Cambodia	8.1	2.48	10.8	0.0	0.36
Cameroon	11.0	2.80	42.4	3.6	5.86
Canada	26.5	1.01	75.9	2.5	400.69
Cape Verde	0.4	2.36	5.3		
Central African Republic	2.8	2.42	42.4	6.3	0.36
Chad	5.5	2.44	27.0	0.1	0.36
Chile	12.8	1.52	83.6	17.1	25.66
China	1110.8	1.18	20.6	0.2	2092.55
Colombia	31.2	2.05	67.4	4.7	48.02
Comoros	0.5	3.08	25.2		
Congo	1.9	2.73	39.5	4.0	1.83
Costa Rica	2.9	2.44	49.8	8.9	2.20
Cuba	10.47	0.98	71.8	2.6	31.89
Cyprus	0.7	1.01	49.5	0.0	3.66
Czechoslovakia	15.8	0.32	65.3	14.4	234.25
Denmark	5.1	0.01	88.9	3.0	60.85

Sources: Col. 1: UN *World Population Prospects* (1988); Col. 2: UN *World Population Prospects* (1988); Col. 3: UN *Prospects of World Urbanization* (1988); Col. 4: World Resources Institute (1988); Col. 5: Oak Ridge National Laboratory; Col. 6: FAO *Fertilizer Yearbook* (1988); Col. 7: *World Motor Vehicle Data* (1)

6 Artificial fertilizer use 1987 kgs per hectare	7 Cars 1986-87 thousands	8 CITES signatories as of Jan.1990	9 Montreal Protocol signatories as of Dec.1989 □ non-ratifying	10 National Green Parties mid-1980s	11 Offices as of Jan.1990 Friends of the Earth Greenpeace W W F
9.7	33.0	●			
133.1					
38.0	589.0	●			
2.9	58.0				
	11.0				
4.5	4,060.0	●	□		G F
28.6	7,073.0	●	■		G F W
197.8	2,685.0	●	■	GP	G F W
50.0	68.0	●			
750.0	60.0				
77.0	45.0	●			F
93.9	35.0				
509.8	3,457.0	●	■	GP	G F W
81.8	0.5	●			
4.9	22.0	●			
	18.0				
1.0					
3.4	40.0	●			
<1.0	17.0	●			
53.2	10,827.0	●		GP	F
200.0	60.0				
180.4	1,100.0				
5.4	22.0		●		
2.0	7.5	●			
<1.0					
7.1	68.0	●	■		
48.1	11,500.0	●	■	GP	G F W
<1.0	43.0	●			
1.7	7.0	●			
52.3	591.0	●	□		
236.5	995.0	●			
94.8	579.0	●			
2.5	26.0	●	□		
180.6	114.0	●			G
197.0					
126.8	131.5	●			F
303.1	2,724.0				
233.5	1,588.0	●	■	GP	G F W

8: US State Department; **Col. 9**: US Environmental Protection Agency; **Col.10**: Sara Parkin, *Green Parties: An International Guide*, London: Heretic ks 1989; **Col. 11**: Friends of the Earth, London office; Greenpeace, Washington office; World Wildlife Fund, Washington office.

	1 Population 1989 millions	2 Population growth rate 1985-90 percentages	3 Urban population 1985 percentages of total population	4 Land area protected 1988 percentages of total land	5 Carbon diox‹ emission‹ 1987 million ton‹
Djibouti	0.4	3.32	77.7	0.0	0.36°
Dominica					0.0
Dominican Republic	6.8	2.21	55.7	11.8	6.59‹
Ecuador	10.5	2.79	52.3	38.4	13.93
Egypt	51.4	2.27	46.4	0.6	70.38
El Salvador	6.3	3.10	39.1	1.0	1.83‹
Equatorial Guinea	0.4	2.31	59.7	0.0	0.0
Ethiopia	48.7	2.79	11.6	1.6	2.56
Fiji	0.7	1.59	41.2	0.3	0.73
Finland	5.0	0.30	84.0	2.6	52.42
France	55.3	0.31	73.4	8.7	335.43
French Guiana			72.6		0.36
Gabon	1.2	2.01	40.9	6.8	5.13‹
Gambia	0.7	2.13	20.1	0.0	0.0
East Germany	16.9	0.15	77.0	0.7	324.07
West Germany	60.4	0.18	85.5	11.3	652.91
Ghana	15.5	3.36	31.5	5.1	2.93
Greece	10.0	0.41	60.1	1.1	52.42
Greenland (Den)			76.8		0.36
Guadeloupe (Fr)			45.7		0.73‹
Guatemala	8.9	2.88	40.0	0.9	2.56
Guinea	6.7	2.48	22.2	0.1	1.1C
Guinea-Bissau	1.0	2.08	27.1	0.0	0.0
Guyana	1.0	1.74	32.2	0.1	1.1C
Haiti	7.3	2.62	27.2	0.4	0.73
Honduras	5.0	3.10	40.0	3.2	1.46
Hong Kong (Br)			92.4		26.39
Hungary	10.7	0.07	56.2	4.8	74.42
Iceland	0.3	0.95	89.4	7.9	1.83
India	813.4	1.72	25.5	4.3	542.20
Indonesia	178.5	1.74	25.3	7.5	120.97
Iran	49.9	2.77	51.9	1.9	139.30
Iraq	18.2	3.31	70.6	0.0	42.89
Ireland	3.8	1.26	57.0	0.3	27.49
Israel	4.5	1.65	90.3	1.7	28.59
Italy	57.5	0.09	67.4	2.5	355.96
Ivory Coast	11.3	3.45	42.0	6.2	4.76
Jamaica	2.5	1.52	53.8	0.0	5.86
Japan	123.3	0.51	76.5	6.1	885.3‹
Jordan	4.1	3.99	64.4	0.4	8.7‹
Kenya	24.4	4.20	19.7	5.4	4.3‹

Sources: Col. 1: UN *World Population Prospects* (1988); **Col. 2**: UN *World Population Prospects* (1988); **Col. 3**: UN *Prospects of World Urbanization* (1988)
Col. 4: World Resources Institute (1988); **Col. 5**: Oak Ridge National Laboratory; **Col. 6**: FAO *Fertilizer Yearbook* (1988); **Col. 7**: *World Motor Vehicle Data* (

6 Artificial fertilizer use 1987 kgs per hectare	7 Cars 1986-87 thousands	8 CITES signatories as of Jan.1990	9 Montreal Protocol signatories as of Dec.1989 □ non-ratifying	10 National Green Parties mid-1980s	11 Offices as of Jan.1990 Friends of the Earth Greenpeace W W F
	15.0				
176.5	3.0				
55.6	103.0	●			
32.1	257.0	●			F
350.5	417.0	●	■		
126.2	73.0	●			
	4.5				
3.9	41.0	●			
89.6	31.0		■		
216.3	1,699.0	●	■	GP	G W
299.0	21,950.0	●	■	GP	G F W
166.7					
3.5	16.0	●			
15.3	5.0	●	■		
337.3	3,462.0	●	■	GP	G F
420.7	28,304.0	●	■	GP	G F W
3.8	60.0	●	■		F
170.6	1,433.0		■	GP	
142.9					
67.9	175.0	●	■		
<1.0	12.0	●			
<1.0	3.0				
26.9	33.0	●			
2.5	35.0				
23.1	41.0	●			
	184.0				F W
259.5	1,660.0	●	■		
2917.1	113.0		■	GP	
53.6	1,471.0	●			W
106.8	1,060.0	●	□		F
65.8	1,600.0	●			
39.7	376.0				
681.5	737.0		■	GP	G F
223.7	649.0	●	□		
190.1	22,800.0	●	■	GP	G F W
6.0	164.0				
91.4	93.0				
432.7	29,478.0	●	■	GP	G F W
34.8	136.0	●	■		
40.6	126.0	●	■		

8: US State Department; **Col. 9**: US Environmental Protection Agency; **Col.10**: Sara Parkin, *Green Parties: An International Guide*, London: Heretic ks, 1989; **Col. 11**: Friends of the Earth, London office; Greenpeace, Washington office; World Wildlife Fund, Washington office.

	1 Population 1989 millions	2 Population growth rate 1985-90 percentages	3 Urban population 1985 percentages of total population	4 Land area protected 1988 percentages of total land	5 Carbon dio emission 1987 million ton
Kiribati			33.5		
North Korea	22.4	2.36	63.8	0.0	143.70
South Korea	44.1	1.89	65.3	5.7	162.77
Kuwait	2.1	4.15	93.5	0.0	32.99
Laos	4.5	2.43	15.9	0.0	0.0
Lebanon	2.9	2.13	80.1	0.0	8.06
Lesotho	1.7	2.61	16.7	0.2	
Liberia	2.5	3.25	39.5	1.4	0.73
Libya	4.2	3.67	64.5	0.1	24.92
Luxembourg	0.4	0.10	81.0	44.2	8.43
Madagascar	11.2	2.90	21.8	1.8	0.73
Malawi	7.9	3.32	12.0	11.3	0.36
Malaysia	17.0	2.12	38.2	4.9	40.32
Maldives			20.2		
Mali	9.1	2.94	18.0	0.7	0.36
Malta	0.4	0.66	85.3	0.0	0.0
Mauritania	2.1	3.08	34.6	1.4	3.29
Mauritius	1.1	1.65	42.4	2.0	1.10
Martinique (Fr)			71.1		0.73
Mexico	87.0	2.39	69.6	0.5	287.41
Mongolia	2.1	2.74	50.8	0.2	9.16
Morocco	24.1	2.30	44.8	0.7	18.33
Mozambique	15.5	2.69	19.4	0.0	0.73
Myanmar (Burma)	40.1	1.89	23.9	0.0	5.49
Namibia			51.3		
Nepal	18.1	2.28	7.7	7.1	131.97
Netherlands	14.7	0.34	88.4	4.4	0.73
New Zealand	3.4	0.86	83.7	11.4	20.89
Nicaragua	3.7	3.36	56.6	0.4	2.20
Niger	6.9	3.01	16.2	0.3	0.73
Nigeria	109.4	3.49	23.0	1.1	55.33
Norway	4.2	0.17	72.8	4.1	43.99
Oman	1.4	3.18	8.8	0.3	21.6
Pakistan	109.7	2.23	29.8	9.4	47.77
Panama	2.4	2.07	52.4	10.9	2.56
Papua New Guinea	3.9	2.38	14.3	0.4	2.2
Paraguay	4.1	2.78	44.4	2.8	1.8
Peru	21.8	2.51	67.4	4.2	22.3
Philippines	59.7	2.25	39.6	1.7	34.4
Poland	38.3	0.70	61.0	5.5	460.0
Portugal	10.5	0.64	31.2	4.2	27.8

Sources: Col. 1: UN *World Population Prospects* (1988); **Col. 2**: UN *World Population Prospects* (1988); **Col. 3**: UN *Prospects of World Urbanization* (1988);
Col. 4: World Resources Institute (1988); **Col. 5**: Oak Ridge National Laboratory; **Col. 6**: FAO *Fertilizer Yearbook* (1988); **Col. 7**: *World Motor Vehicle Data* (

INTERNATIONAL TABLE

6 Artificial fertilizer use 1987 kgs per hectare	7 Cars 1986-87 thousands	8 CITES signatories as of Jan.1990	9 Montreal Protocol signatories as of Dec.1989 □ non-ratifying	10 National Green Parties mid-1980s	11 Offices as of Jan.1990 Friends of the Earth Greenpeace W W F
311.7					
422.4	844.0				
82.0	435.0				
<1.0	10.0				
67.1					
12.5	6.0				
7.3		●			
28.8	404.0				
	162.0	●	■	GP	G
5.9	49.0	●			
20.4	16.0	●			
159.6	1,269.0	●	■		F W
			■		
13.9	21.0				
45.9		●	■		
5.0	15.0				
306.9	35.0	●			
1150.0					
75.3	5,403.0		■		
18.4					
35.7	565.0	●	□		
2.1	87.0	●			
15.4	35.0				
22.9		●			
687.9	5,118.0	●	■	GP	G F W
708.6	1,424.0	●	■	GP	G F W
43.3	32.0	●			F
<1.0	23.5	●			
9.9	774.0	●	■		
270.4	1,623.0	●	■	GP	G W
91.7	109.0				
82.9	273.0	●			F W
65.7	104.0	●	■		
38.1	17.0	●			F
4.4	88.5	●			
62.2	390.5	●			
65.0	359.0	●	□		
222.4	4,000.0				F
102.6	1,290.0	●	■		F

8: US State Department; **Col. 9**: US Environmental Protection Agency; **Col. 10**: Sara Parkin, *Green Parties: An International Guide*, London: Heretic
s 1989; **Col. 11**: Friends of the Earth, London office; Greenpeace, Washington office; World Wildlife Fund, Washington office.

	1 Population 1989 millions	2 Population growth rate 1985-90 percentages	3 Urban population 1985 percentages of total population	4 Land area protected 1988 percentages of total land	5 Carbon dio× emission 1987 million ton
Puerto Rico (US)			70.7		14.29
Qatar	0.4	5.44	88.0	0.0	11.36
Romania	23.7	0.68	49.0	0.6	205.66
Rwanda	6.9	3.36	6.2	10.5	0.36
Saudi Arabia	13.5	3.84	72.4	0.2	164.60
Senegal	7.2	2.71	36.4	11.3	1.83
Sierra Leone	3.9	1.93	28.3	1.4	0.73
Singapore	2.7	1.09	100.0	4.3	28.22
Solomon Islands	0.3	3.96	9.7	0.0	
Somalia	5.1	2.11	34.1	0.5	1.10
South Africa	35.8	2.53	55.9	4.7	278.24
Spain	39.5	0.62	75.8	3.4	161.30
Sri Lanka	17.2	1.48	21.1	10.6	3.66
Sudan	24.2	2.89	20.6	0.9	3.29
Suriname	0.4	1.46	45.7	4.6	1.10
Swaziland	0.7	3.14	26.3	2.3	0.36
Sweden	8.3	0.11	83.4	4.2	56.80
Switzerland	6.4	0.04	58.2	3.0	38.12
Syria	12.2	3.69	49.5	0.0	2.20
Taiwan					
Tanzania	26.1	3.65	22.3	12.0	25.66
Thailand	54.8	1.61	19.8	7.8	52.7
Togo	3.3	3.06	22.1	8.5	0.0
Trinidad & Tobago	1.3	1.59	63.9	3.1	17.5
Tunisia	7.7	2.17	56.8	0.4	9.8
Turkey	53.6	2.06	45.9	0.3	122.4
Uganda	17.8	3.49	9.5	6.7	0.7
United Arab Emirates	1.5	3.46	77.8	0.0	50.2
United Kingdom	56.2	0.02	91.5	6.4	569.6
USA	246.3	0.86	73.9	7.4	4457.1
USSR	289.3	0.93	65.6	0.0	3745.5
Uruguay	3.1	0.76	84.6	0.2	2.9
Venezuela	19.2	2.61	86.6	8.4	95.3
Vietnam	64.8	2.05	20.3	0.6	18.3
Western Sahara			52.9		0.3
South Yemen	2.4	3.01	39.9	1.3	2.9
North Yemen	7.7	2.92	20.0	0.8	5.4
Yugoslavia	23.8	0.63	46.3	1.3	120.6
Zaire	33.8	3.04	36.6	3.9	3.2
Zambia	7.6	3.43	49.5	8.6	2.2
Zimbabwe	10.1	3.61	24.6	7.1	15.3

Sources: Col. 1: UN *World Population Prospects* (1988); **Col. 2:** UN *World Population Prospects* (1988); **Col. 3:** UN *Prospects of World Urbanization* (1988
Col. 4: World Resources Institute (1988); **Col. 5:** Oak Ridge National Laboratory; **Col. 6:** FAO *Fertilizer Yearbook* (1988); **Col. 7:** *World Motor Vehicle Data*

INTERNATIONAL TABLE

6 Artificial fertilizer use 1987 kgs per hectare	7 Cars 1986-87 thousands	8 CITES signatories as of Jan.1990	9 Montreal Protocol signatories as of Dec.1989 □ non-ratifying	10 National Green Parties mid-1980s	11 Offices as of Jan.1990 Friends of the Earth Greenpeace W W F
	1,280.0				
172.5	78.0				
130.1	250.0				
2.0	7.0	●			
367.8	2,245.0				
4.0	78.0	●	□		
<1.0	23.0				F
1833.3	236.0	●	■		
4.0	17.0	●			
53.8	3,079.0	●			W
98.9	10,219.0	●	■		G F W
113.1	148.0	●			
2.5	34.0	●			
161.6	35.0	●			
39.6	21.0				
135.7	3,367.0	●	■	GP	G F W
430.6	2,733.0	●	■	GP	G F W
40.5	95.0		■		
	1,255.0				
8.9	42.0	●			F
29.3	572.0	●	■		
7.5	3.5	●	□		
52.5	264.0	●	■		
22.2	171.0	●	■		
61.8	193.0				
<1.0	32.0		■		
163.2	250.0	●			
355.5	20,097.0	●	■	GP	G F W
93.3	137,324.0	●	■		G F W
117.8	11,800.0	●	■		G F
42.7	164.0	●			F
158.0	1,601.0	●	■		
62.9					
12.1	12.0				
5.8	121.0				
132.8	3,040.0			GP	
<1.0	94.0	●			
18.4	98.0	●			
50.4	254.0	●			

8: US State Department; **Col. 9**: US Environmental Protection Agency; **Col. 10**: Sara Parkin, *Green Parties: An International Guide*, London: Heretic ks 1989; **Col. 11**: Friends of the Earth, London office; Greenpeace, Washington office; World Wildlife Fund, Washington office.

COMMENTARY

1 A GLOBAL GREENHOUSE

Certain gases, including carbon dioxide, methane, nitrogen oxides and chlorofluorocarbons (CFCs), are accumulating in the upper atmosphere at a faster rate than ever before. Proponents of the theory of global warming warn that this mixture is creating a layer that traps the earth's radiated heat, which will cause global temperatures to rise significantly. It is argued that this could lead to a rise in sea levels, higher tides and extraordinary storm surges; and the impact of these changes would mean salination of drinking water supplies, disruption of food crop and plant habitats, the expansion of deserts and the inundation of entire islands. Low-lying countries such as Bangladesh, the Netherlands, the Maldives and many small Pacific island states are especially vulnerable. Most of the world's major cities, including many capitals, are coastal and even a slight rise in sea-level would bring some threat of flooding. At a more extreme level, large-scale changes to weather and agricultural patterns would wreak havoc with local livelihoods and national economies.

Environmentalists are mapping out strategies to stave off global warming. These focus on the need for cutbacks in the use of fossil fuels (see Maps 23 and 31) and CFCs (see Map 28), as well as the need to call a halt to further global deforestation (see Maps 5 and 35). Such an agenda has sweeping implications. It would require dramatic changes in the priorities of industrial economies and industrial lifestyles; more difficult still is its potential impact on the plans of poor countries for economic expansion. The enormity of the demands is paralysing political and economic agreements and creating an uproar in scientific circles. But business and governments have short-term vested interests which make it convenient to dispute the theory of global warming.

Sources for map and commentary:
Boyle, Stewart & John Ardill, *The Greenhouse Effect*, London: New English Library, 1989; Hansen, J. et al, 'Global Climate Changes as Forecast by the Goddard Institute for Space Studies'. *Journal of Geophysical Research*, vol. 93, 1988; *Climatic Change, Rising Sea Level and the British Coast*, Institute of Terrestrial Ecology, NERC, 1989; Stevens, William, 'Adapting to Global Greenhouse', *International Herald Tribune*, November 16, 1989; World Resources Institute/International Institute for Environment and Development, *World Resources 1988-89*, New York: Basic Books, 1988.

2 A POPULAR PLANET

In 1985, Africa and Europe were approximately the same size in population; by 2025, Africa's population will be three times larger than that of Europe. Nigeria alone will grow from being half the population size of the USA to being slightly larger. In the mid-1980s, the Middle East dominated the ranks of states with the highest population growth rates: Oman, Libya, UAE, Qatar, Saudi Arabia, Kuwait, Bahrain, and Iran were all in the top ten. Dramatic increases such as these, together with the shifting relationships within the global population balance, form the backdrop to many of the environmental themes in this atlas.

But the relationship between population and the environment is complex. Some would argue, including some environmentalists, that there is a direct correlation between the world's growing population and the world's growing environmental problems. The evidence does not bear this out. While population growth is now a Third World phenomenon, the large-scale global environmental crises we face are largely the product of the voracious resource appetites of industrial economies and lifestyles. In terms of how much we burden the atmosphere, one economist estimates that the 'cost' of one US citizen is 16 times that of a Third World person; of a West European, five times. The average person in the developed world consumes 10 times as much energy, 10 times as much steel, 15 times as much paper, and one-and-a-half times as much food as does the average person living in the Third World. If we check off the world's most pressing environmental problems – fossil fuel pollution (and the related threat of global warming), the use and production of ozone depleting chemicals, the resource use and pollution generated by oil dependency, acid rain emissions, production of household waste, municipal waste, toxic industrial waste, the luxury trade in exotic animal and bird species, and the stripping of the world's forests – it becomes clear that the biggest **99** contributing cause of these problems is not the 'excessive' number of people in the poor countries of the world, but the excessive pressures exerted by the richer countries.

This is not to suggest that growing numbers of people do not burden natural ecosystems; they do. Destruction of wildlife habitats, forest cutback for subsistence agriculture and deterioration in the quality of urban water, to name but a few pressing environmental problems, stem in some measure from the sheer pressure of human need. Ultimately, though, most of these are problems of poverty and misdirected public priorities rather than problems of numbers. Military priorities are often the biggest stumbling block. For the cost of a single British Aerospace Hawk aircraft, for example, 1.5 million people in the Third World could have clean water for life; the annual cost of the proposed anti-desertification programme for Ethiopia is the equivalent of two months of Ethiopian military spending.

In turn, environmental problems themselves, in a world of social inequities, encourage high rates of population growth - in countries where infants die regularly from gastro-intestinal disease brought on by lack of clean water, for example, women have to bear higher numbers of children simply to meet perceived population replacement levels.

Population growth may well be a 'problem', but not in the ways that most of us are encouraged to believe. And the solution to environmental problems is not simply, or even primarily, through population controls; indeed, the reverse may well be true.

Sources for map and commentary:
Dankelman, Irene & Joan Davidson, *Women and Environment in the Third World*, London: IUCN/Earthscan, 1988; Greanville, Patrice, 'Who's Got the Answer to the Overpopulation Riddle?' *Animals' Agenda*, November 1988; United Nations, Department of International Economic and Social Affairs, *World Population Prospects, 1988*, 1989; World Commission on Environment and Development, *Our Common Future*, Oxford: Oxford University Press, 1987.

3 A SHRINKING WORLD

Overexploitation, loss of habitat (particularly through deforestation), hunting and the luxury trade in animals are the major causes of the spiralling scale of animal extinctions. By the year 2000, between 40,000 to 50,000 species will be lost annually.

Information about the world's animal communities is very uneven. As with plant species, more is generally known for richer, industrialized countries. Many others, including a number rich in both animal and plant species such as Brazil or Zaire, have little in the way of scientific monitoring services. The data is uneven in other important respects. More money is available to fund studies of large mammals and primates (perhaps because of perceived kinship and 'usefulness' to humans). Little attention is given to less attractive animals and birds and even less to fish or insects. There are only broad estimates for losses of fish and although many insect species are disappearing there is not even a 'best estimate' for the number becoming extinct.

Birds are especially vulnerable to extinction because they are migratory and range over a wide territory. There needs to be coordinated international action for the protection of birds but this will be difficult to achieve. For example, the annual hunting massacre of millions of songbirds along their migration paths in the Mediterranean has gained considerable notoriety and drawn protests from the EEC. It has not yet been halted. A third of Europe's birds are threatened with extinction, of which hunting is major cause, followed by habitat destruction. Hunters kill an estimated 900 million migrating birds each year. They justify this slaughter as a male ritual with links to cultural traditions that go back thousands of years.

Environmentalist Norman Myers comments wryly on the path down which we seem to be rushing: 'Unwittingly for the most part, but right around the world, we are eliminating the panoply of life. We elbow species off the planet, we deny room to entire communities of nature, we domesticate the Earth. With growing energy and ingenuity, we surpass ourselves time and again to exert dominion over fowl and fish ... Eventually we may achieve our aim, by eliminating every other competitor for living space on the crowded Earth. When the last creature is accounted for, we shall have made ourselves masters of all creation. We shall look around and we shall see nothing but each other. Alone at last.'

Sources for map and commentary:
Cocker, Mark, 'The Fall of Europe's Birds', *The Guardian* (London) July 25, 1989; IUCN Conservation Monitoring Centre, *Red List of Threatened Animals*, Cambridge: IUCN, 1988; Myers, Norman, ed., Gaia: *An Atlas of Planet Management*, London: Pan; New York: Doubleday, 1984; Organization for Economic Cooperation and Development (OECD), *OECD Environmental Data Compendium, 1989*, Paris: OECD, 1989; World Resources Institute/International Institute for Environment and Development, *World Resources 1988-89*, New York: Basic Books, 1988.

4 CUTTING OUR ROOTS

All life on the planet depends, directly or indirectly, on plants. And yet plants and plant species are being destroyed at an alarming rate. If present rates of habitat destruction continue, then some 40,000 - 60,000 vascular plants will face extinction within the next fifty years. In this map, we have included as much information as possible on small islands because they have distinctive biomes, tend to be rich in plant species and are suffering high rates of loss.

Beyond broad assessments, detailed information about plant extinctions is patchy. The Threatened Plants Unit of the International Union for Conservation of Nature (IUCN) in the UK collects the only available international database on plants, and publishes a floristic inventory in their Red Data Books. To date, Red Books have been compiled for 70 countries and islands. Most countries of the North, with predominantly temperate vegetation, are well covered by IUCN data, but there are few Red Books for countries of the South, where the vegetation is mostly tropical - and where most of the world's plants grow. There is almost no plant inventory or assessment for key areas such as Brazil, much of Central America and most of Africa. In the case of plant ecology, what we do not know is more problematic than what we do know. In general, it must be assumed that low rates of extinction reflect a lack of data more than they reflect the reality of plant losses.

The destruction of plants will threaten the world's food supplies and undermine medical advances. Wild plants are being lost which are needed to protect and maintain the world's major food crops. Eighty percent of our food comes from these crops and they are already suffering a dramatic decline in genetic diversity (see Map 10). Four-fifths of the world's population depends on natural plant sources for medicines. Even high-technology Western medicine uses a very high proportion of life-saving drugs that are initially derived from plant sources and then synthesized for large-scale production.

Sources for map and commentary:
Nature Conservancy Council, 'British Rare Plants: Threats and International Responsibilities', Data Support for Education, Sheet 60, 1989, Peterborough (UK) 1989; Davis, S.D., et al, *Plants in Danger: What do we Know?* Cambridge: IUCN, 1986; Fagan, Mary, 'Food Supply at Risk from Extinction of Plant Species', *The Independent* (London) December 16, 1989; World Resources Institute/International Institute for Environment and Development, *World Resources 1988-89*, New York: Basic Books, 1988.

5 RAINFOREST

Tropical forest originally covered about 16 million sq. kms. throughout the world; today, probably less than half remains. Although tropical forest (including rainforest) accounts for only a third of the world's total forest, it contains four-fifths of the world's vegetation: a single hectare of primary forest may support plant material weighing anywhere from 300 to 500 tonnes. More than that, it provides the habitat for about 50 percent of all animal and bird species. As tropical forest is destroyed, so other plant and animal life will disappear too. The recent recognition that tropical forests influence the global climate, and in particular may slow down the effects of global warming, has raised the stakes around tropical forest protection. However, to put an end to tropical deforestation will require strict regulation of multinational agricultural and logging industries. Although sustained international cooperation is essential, as well as a radical shift in the primary-resource economies of many countries, as of early 1990 there is still no international coordinating body.

The geography of rainforest has never been precise and reliable information on rates of deforestation does not exist. There are both practical and technical difficulties. One of these is the varying definitions of tropical forest, such as lowland, montane, dry, rain, swamp and so on. Further, when measuring deforestation, it is difficult to determine the difference between 'degraded' forest land and totally destroyed forest, especially since the one usually leads to the other. The term 'deforestation' itself means different things to different people. To commercial interests, deforestation means the loss of commercial timber stock. To environmentalists, deforestation is the destruction of an entire forest ecosystem, even though stands of trees may remain.

Whatever the definition, at the current rate of deforestation virtually all remaining tropical forests will disappear within 40 years. The tropical forests of the Ivory Coast, Nigeria, India, and Thailand are already severely depleted. By 2010, there will be very little forest left in Asia, and East and West Africa and almost none in Central America. Even in South America, the only large stands will remain in western Brazil and the highlands of Guyana.

Rainforest destruction has highlighted an important issue of social justice: the land rights of indigenous peoples. Large numbers of tribal people are each year losing their lands in the name of development and progress, and in the heart of Borneo, Brazil, the Philippines, and Malaysia, tribal peoples have been fighting a virtual life and death struggle to protect their rainforest homelands. Indigenous peoples and environmentalists are forming alliances to protect them.

There are many widely-accepted causes of rainforest loss: logging by commercial timber companies (see Map 35); migration of poor agricultural families into forests, often as part of government-promoted relocation schemes; conversion of forest land into commercial cash cropland, usually to grow beef or soya, and often to raise capital to pay off foreign debts; and firewood collection and the production of charcoal for energy (see Map 21).

Sources for map and commentary:
Campbell, David & David Hammond, *Floristic Inventory of Tropical Countries*, London: Friends of the Earth, 1989; Myers, Norman & Richard A. Houghton, 'Deforestation Rates in Tropical Forests, and Their Climatic Implications' (First Draft), London: Friends of the Earth, 1989; 'Paradise Lost?' London: EarthLife Foundation, 1986; World Resources Institute/International Institute for Environment and Development, *World Resources 1988-89*, New York: Basic Books, 1988.

6 DRINKING WATER

The World Health Organization defines reasonable access to safe drinking water in an urban area as 'access to piped water or a public standpipe within 200 metres of a dwelling'; for rural areas, the WHO definition of reasonable access is 'drinking water within 15 minutes walking distance'. Interpretation of these definitions, even in the definition of what is rural and what is urban, is clearly open to subjective judgement; variations in definition mean that comparisons between countries may be misleading. In general, much of the international data available on water provision must be accepted with scepticism.

In 1980, the United Nations launched the International Drinking Water Supply and Sanitation Decade in an effort to improve access to potable water and sanitation in developing countries. The mid-decade report showed promising signs for rural dwellers: worldwide, access to safe water had increased from 33 percent to 45 percent; the picture for the world's urban population had remained static: the worldwide average being about 75 percent. However, the report also noted that the per capita cost of providing water and sanitation services continued to increase despite the development of less expensive technologies, which suggests that governments may not be able to sustain improvements in water supply.

To support a reasonable quality of life requires approximately 80 litres of water per person per day. Around the world, average consumption ranges from 5.4 litres per day in Madagascar to more than 500 litres (approximately 160 US gallons) per day in the USA. Everywhere, demands on water are high, and growing fast. In the USA, it is estimated that household water consumption will quadruple between 1967 and 2000, and that the demands of industry will increase fivefold.

The main problems in relation to water supply vary between rich and poor countries, although sewage is polluting rivers everywhere (see Map 16). In poor countries, the greatest difficulty is the cost of bringing piped water to more people. In the rich world, there are growing problems of protecting water supplies from pollution. Groundwater and rivers are threatened by the use of pesticides and fertilizers in agriculture and by disposal methods for the increasing quantities of industrial and toxic waste produced by industry (see Maps 25 and 27). The UK Department of the Environment reported in 1988 that some drinking water sources had had to be 'abandoned because of pollution from landfill sites'. And in February 1990 the *Observer* reported that over 1300 landfill sites used for disposal of toxic waste were posing 'some risk' to water supplies.

Ideally, drinking water should come from unpolluted groundwater. However, this is an ideal in rapid retreat, and currently more than half the population of OECD countries and 70 percent of the US population drinks water that has been passed through waste water treatment plants. The danger in this is that not all waste can be removed in treatment, and the chemical agents used in treatment may themselves pose health risks.

Sources for maps and commentary:
Goldsmith, Edward & Nicholas Hildyard, eds., *The Earth Report*, London: Mitchell Beazley, 1988; Myers, Norman, ed., *Gaia: An Atlas of Planet Management*, London: Pan; New York: Doubleday, 1984; Lean, Geoffrey and Polly Ghazi, 'Britain's Buried Poison', *The Observer*, February 4, 1990; Organization for Economic Cooperation and Development (OECD), *OECD Environmental Data Compendium, 1989*, Paris: OECD, 1989; World Health Organization (WHO), International Drinking Supply and Sanitation Decade, *Review of National Progress 1983*, Geneva, 1986; WHO, *World Health Statistics Annual*, Geneva, 1986.

7 HUNGER AND PLENTY

The Hunger Report, an annual assessment of the state of hunger in the world, introduces the issue in this way: 'No one really knows how many hungry people there are in the world. No one knows the toll of hunger because hunger is difficult to define, because the statistical data are weak or nonexistent, and because efforts to improve data collection and analysis have been limited. But beyond these real difficulties, we may not know how many hungry people there are in the world because we may not want to know. To know how many hungry there are in a world of plenty is to measure the inadequacies of our economies to sustain all, of our societies to provide for all, and of our common humanity to care for all'.

It is by now commonplace to observe that there does exist more than enough food in the world to feed everyone adequately. Hunger reflects social derangement more than it reflects the capacity for food production.

Most international statistics on food and hunger measure the average availability of food supply, which is not the same as food consumption: because food is 'available' does not mean that everyone has access to it. Food consumption is mediated by class, gender, and age, among other variables. It is subject to serious disruption, most noticeably by war. Food supplies are often the intentional targets of armed conflict, which, not surprisingly, affects the poorest, hungriest, and most vulnerable in the population first, chief among whom are women and their dependent children.

The production and supply of food are obviously affected by environmental factors. Agendas for action on food and hunger have to tackle such issues as land erosion (see Map 11); increasing agro-chemical dependence (see Map 9); the need to conserve wild genes for the protection and maintenance of crops (see Map 10); and even the possibility of a changing world climate.

Sources for map and commentary:
Food and Agriculture Organization (FAO), *The State of Food and Agriculture*, 1989; FAO, *Production Statistics Yearbook*, Rome, 1987; Kates, Robert, et al, *The Hunger Report: 1988* and *Update: 1989*, Brown University: Alan Shawn Feinstein World Hunger Program, 1989.

8 LIVING OFF THE LAND

The proportion of people directly engaged in agriculture worldwide is dropping rapidly: in developing countries, from 70 percent in 1970 to 50 percent in 1990; in developed countries over the same period, from 16 percent to 8 percent. Meanwhile the number of people that each hectare of land must support is increasing dramatically. The amount of land available is relatively fixed, while the world's population is growing.

The race to obtain bigger crop yields out of a fixed agricultural base has been directed by large multinational chemical and agro-business enterprises. While all evidence shows that small farms are more efficient and productive than large farms, the worldwide trend is towards larger and larger consolidated agricultural holdings. Similarly, although food grown with heavy doses of chemicals is known to have a lower nutritional value, the 'drugging' of world agriculture continues (see Map 9). Finally, although land reform has proven to be the single most-effective means of increasing agricultural productivity, and despite recent revolutions waged on the promise of land reform, little is changing: in Peru, for example, 93% of all agricultural land is owned by the top 10% of landowners; in El Salvador, 78%; in Guatemala, 76%.

Much deforestation in Central and South America (see Map 5) is a direct consequence of distorted land tenure systems: huge areas of forest have single owners, who then assume the right to clear-cut forest to make way for large cattle-ranching enterprises. The cattle businesses themselves are often owned by foreign companies in conjunction with a country's largest landholders. Intervention in deforestation almost always brings environmentalists into direct conflict with the political and economic elite who own most of the land. In many countries, land reform and environmental reform must go hand in hand.

Sources for map and commentary:
Gramfors, Bo, *Concise Earth Atlas: The Book of Everything*, Stockholm: Esselte Map Service, 1989; Kurian, George, *The New Book of World Rankings*, New York: Facts on File, 1984; Food and Agriculture Organization (FAO), *Production Statistics Yearbook, 1988*, Rome: 1989.

9 CHEMICAL FIX

World agriculture is hooked on synthetic chemicals. The use of agro-chemicals has increased tenfold since World War II, and most crops are now grown with the assistance of artificial fertilizers and biocides. In 1988, for example, American farmers used almost 19 million tonnes of synthetic fertilizers and 255,000 tonnes of pesticides. The demand is for new and stronger chemicals, and for heavier doses. But chemical dependence is cyclical: fertilizers, which may appear to be the most benign of agro-chemicals, disrupt natural soil balance and crop ecology; this then encourages the use of herbicides, insecticides, and pesticides; these chemicals then strip soils of nutrients, which then necessitates heavier use of fertilizers.

Agricultural chemicals pose a grave pollution threat to entire ecosystems. Chemical runoff from croplands causes river eutrophication; insecticides and herbicides kill 'benign' animals and birds as well as 'pests'. DDT nearly wiped out dozens of domestic bird species on the eastern seaboard of the USA before its use was banned in the early 1970s; meanwhile in Thailand, DDT is now threatening drinking water supplies throughout the country; and over large regions of Central America, where pesticide use levels are among the highest in the world, wildlife is being decimated and the natural ecology is crumbling.

In addition to the indirect health threat of chemical residues remaining in food, agro-chemicals pose additional direct threats to human health. One person in the world dies from pesticide poisoning every two hours. In the USA, the Environmental Protection Agency estimates that at least 66 of the roughly 300 pesticide ingredients commonly used by farmers are 'probable carcinogens', and dozens more are known to cause birth defects, nervous system disorders, and other chronic illnesses. Accidents in the manufacture and storage of agro-chemicals have killed thousands of people worldwide and caused grave environmental damage. Many of the major industrial accidents over the past twenty years have involved agro-chemical production facilities: among the most prominent of these have been the accidents at Bhopal and Seveso plus the Basle chemical spill into the River Rhine (see Map 26).

Although agricultural chemicals are a known threat to the environment and to human health, the agro-chemical industry is very big business and the chemical lobby is powerful enough to evade or defeat most regulatory efforts. One favoured tactic of the industry is to shift production and markets when domestic regulation threatens chemical production. Current estimates suggest that 75 per cent of the pesticides used in the Third World would not be allowed in the USA, and many of the chemicals used by farmers in poorer countries are actually banned in first world markets, but produced by American and West European firms for overseas use. The USA exports more than 18 million kgs of DDT annually, but a clear example of the shift from home consumption to export is offered by British export figures for DDT. In 1980, British exports were 1840 kgs, yet in 1984, after the pesticide was banned for domestic use, exports rose to 125,503 kgs.

Fertilizer use is monitored closely by the United Nations' Food and Agriculture Organization (FAO). There is no comparable record of international pesticide manufacture and use; the best monitoring of this side of the chemical industry comes from independent environmental and consumer groups.

Sources for map and commentary:
Food and Agriculture Organization (FAO), *Fertilizer Yearbook 1988*, Rome, 1989; FAO, *Production Statistics Yearbook 1987*, Rome 1988; FAO, *Trade Statistics Yearbook 1988*, Rome 1989; Goldsmith, Edward & Nicholas Hildyard, eds., *The Earth Report*, London: Mitchell Beazley, 1988; Mercier, Michel & Morrell Draper, 'Chemical Safety: The International Outlook', *World Health*, August/Sept., 1984; Schneider, Keith, 'Pesticide Regulation: Slow and Unsteady,' *New York Times*, March 19, 1989; Weir, David & Mark Schapiro, *Circle of Poison*, Berkeley, CA: Institute for Food and Development Policy, 1981.

10 SHRINKING GENES

Genetic diversity is essential for the maintenance of ecological stability. Food plants that people have 'domesticated' do not endure on their own, isolated as they are from cross-fertilization with other genetically diverse plants. Most of the world's food now derives from about 20 plant species. Yet food crops must always be reinforced with infusions of genes from wild species. These wild relatives possess genes, for example, that protect against pests and disease, improve tolerance to drought and even improve nutritional values.

A number of cultivated crops have wild gene relatives found only in rainforest, which is being rapidly destroyed (see Map 5). The loss of this and other important plant habitat is one primary problem. So also is the spread of monoculture practised by large-scale agro-business, especially in industrial countries. This reliance on monoculture is itself destroying important wild plant genetic material. One observer has noted: 'The products of agro-technology are displacing the source upon which the technology is based. It is analogous to taking stones from the foundation to repair the roof.' The foundation of plant genetic diversity in the world is on the point of collapse.

Genetic material, or germplasm, can be collected and stored as seeds, plants, cuttings, pollen and/or tissue culture, depending on the species. Most of the critical areas for collection of wild genetic material so far identified are in the 'Third World'. However there are many less valuable sources that also need urgent attention and preservation. The identification of the genetic origins of plants is a herculean task, and one barely begun.

There is no such thing as self-sufficiency in plant resources. The International Board for Plant Genetic Resources (IBPGR) is supported by the United Nations and has taken global responsibility for the coordination of germplasm collection and storage. Most species now being collected and preserved by the IBPGR as priorities are cultivated crops. A number of international, mostly private, agencies are also engaged in collecting and preserving genetic material in gene banks.

Genetic engineering involves the deliberate manipulation of an organism's nucleic acid, or genes, to give it new properties. Genetic engineering has great potential for improving food production. **105**

increasing resistance to disease and replacing 'defective genes'. However, it is possible for a genetically-engineered organism to escape the confines of a controlled experiment and become an opportunistic organism that could cause widespread disease and ecological disruption. The deliberate release of genetically-engineered organisms in field trials has been vigorously opposed, but releases are occurring at an increasing rate. Four releases of engineered organisms have occurred in countries other than those identified on the map.

Sources for map and commentary:
UK Department of the Environment, personal communication; Myers, Norman, ed., *Gaia: An Atlas of Planet Management*, London: Pan; New York: Doubleday, 1984; Prescott-Allen, R & C. Prescott-Allen, *Genes from the Wild*, London: Earthscan, 1988.

11 LOSING GROUND

Desertification was a big media issue of the 1970s and early 1980s. Described as a problem of over-exploitation of marginal lands in arid climates (especially in the Sahel area of Africa), the term conjured up images of 'deserts on the march'. Desertification has now taken on a low profile, replaced by newer global concerns such as ozone depletion and the greenhouse effect, and the concept itself has come under much critical review.

Nonetheless, soil loss - whether technically described as desertification or not - is still a pressing global concern. Soil loss is threatening agricultural viability on all six continents. In Australia, six tonnes of topsoil are eroded for every tonne of produce grown; in the USA, a third of all cropland is affected by erosion; over much of Asia, 40 percent of the land is at high risk of erosion or desertification.

Erosion - the removal of soil by wind or water - is a natural and ongoing process, but one that is dramatically aggravated and advanced by human activity, particularly the removal of vegetation. When forests are cut down or land is cleared for agriculture or animal grazing, there is nothing left to stabilize the soil, and rapid soil loss follows. In the late 1980s, the deadly landslides and floods that have plagued many Himalayan states, Bangladesh in particular, are largely attributed to the linked problems of deforestation and erosion.

Traditional peasant farming systems are often blamed for erosion problems. In fact, the greater problem is the spread of mechanized farming, the development of cash crops for export, and the introduction of vast livestock ranges that have led to overgrazing and unsustainably intensive agriculture.

Sources for map and commentary:
Goldsmith, Edward & Nicholas Hildyard, eds., *The Earth Report*, London: Mitchell Beazley, 1988; Myers, Norman, ed., *Gaia: An Atlas of Planet Management*, London: Pan; New York: Doubleday, 1984; United Nations Environment Programme (UNEP) *World Conservation Strategy*, 1980; World Resources Institute/Institute for Environment and Development, *World Resources 1988-89*, New York: Basic Books, 1988.

12 FISHING FOR FOOD

Fish is a major source of food for many in the world - fisheries supply 23 percent of the world's protein - and fishing provides a livelihood for millions of people. Since 1950, global yields of fish (both marine and fresh-water) have risen over fourfold, from 20 million tonnes to more than 85 million tonnes.

However, world fish stocks are now threatened by many factors, including over-exploitation, pollution, (see, for example, Map 22) and wetlands destruction (see Map 14). After some years of increasing yields, the stocks of many fisheries are declining precipitously. According to a mid-1980s survey by the United Nations' Food and Agriculture Organization (FAO), 11 major ocean fishing grounds, including some newly exploited grounds, are now overfished to the point of collapse. Given the greater dependence on fish for food in Third World countries, it is the poor who will be suffering first and most as fisheries decline.

Statistics monitoring the state of fisheries are often contradictory, and usually considerably out of date. Estimates of catches are highly unreliable and even the most reliable figures, from FAO, must be treated with some caution. In most statistical summaries, 'artisanal' fishing and small coastal independent fishing enterprises are often overlooked, although these supply much of the demand for fish in poorer countries.

Sources for map and commentary:
Food and Agriculture Organization (FAO), *World Food Report*, Rome, 1985; FAO, *Fishery Statistics*, 1986, 1987; Myers, Norman, ed., *Gaia: An Atlas of Planet Management*, London: Pan; New York: Doubleday, 1984.

13 CITY SPRAWL

Cities have always acted as focal points in the development of nation-states, and as symbols of opportunity in the popular imagination. They remain symbolically powerful, but the reality of urbanization is rapidly changing. Some cities remain small in scale and functionally intact. But as other cities — the 'hotspots' on this map, as well as others unnamed — grow into vast conglomerations, they are both uncontrollable and unsustainable. Governments cannot afford to provide the vast infrastructure that is needed to accommodate adequately the huge numbers of people drawn to city life.

In consequence, the social structure of 'mega-cities' around the world is increasingly bifurcated into a small, powerful, and often extremely wealthy elite class, and a huge underclass that is ill-housed or homeless, ill-fed, and unserved by even basic municipal services. There is no poverty more squalid than urban poverty, and in the rapidly urbanizing centres of the world, city life and poverty seem synonymous. Most of the people migrating to the largest and fastest-growing cities in the world today will end up living in shanty-towns, slums or 'cardboard cities' on the outskirts, with no running water, sanitation services or health services. Many will end up with work as street vendors, as exploitatively-paid piece-workers or, especially many young women, as prostitutes or domestic servants for the wealthy.

Many of today's new urban arrivals are unwilling migrants, people displaced from their rural landholdings by war, famine, massive 'development' schemes such as large dams (see Map 20), and agricultural 'modernization'. For others, willing migrants, the attractions of the city are not diminished by the reality of urban poverty: cities still offer the possibility of economic opportunities not found elsewhere, the hope of greater social, political or religious freedom, and amenities not possible in rural settings.

The locus of global urbanization is shifting. In 1960, six out of ten of the world's largest cities were in industrialized countries: New York, Los Angeles, London, the Rhine/Ruhr conglomeration, Paris, and Moscow. By the year 2000, only two cities in the industrial world, New York and Tokyo, will remain in the top ten ranking. Mexico City, ranked 15 in size in 1960, will be the world's largest city by the year 2000.

As cities grow, in absolute size and proportional importance, they play a larger part in shaping the global environment. Under the pressure of urban sprawl, agricultural land is paved over, wetlands are drained, natural habitats are destroyed, rivers are diverted, and wildlife habitats are pushed further and further into the margins. The internal ecology of cities also changes: pollutants of all kinds are concentrated in urban centres, and many cities have become highly hazardous places to live. Environmental action must acknowledge the global nature of the urban problem. Efforts must be made to make cities more humane places to live - and more environmentally benign.

Sources for map and commentary:
Atlas de Carreteras Mexico, Mexico City, Guia Roji, 1987; McDowell, Bert, 'Mexico City: An Alarming Giant', *National Geographic*, August 1984; United Nations, Department of International Economic and Social Affairs, *The Prospects of World Urbanization*, 1987; United Nations, *Compendium of Human Settlement Statistics*, 1983.

14 COASTAL CRISIS

Wetlands fringe most coastlines of the world, as mangroves in the tropics and subtropics, and as salt marshes in the temperate zones. They mediate between the ecosystems of land and sea, serving as huge reservoirs of species diversity, providing rich wildlife habitats for birds, animals, and fish. They provide the spawning grounds for most of the world's commercial fisheries, they act as natural buffers that temper the effects of extreme flooding and drought, they filter pollutants out of maritime systems, they recharge groundwater drinking supplies, and they stabilize coastal erosion. They stabilize and support the health of the planet.

Wetlands are under extreme threat almost everywhere in the world. Clear-cutting is perhaps the greatest single threat, but pollution and urban development rank close behind. Urban development alone has claimed almost half the wetlands in the USA, and in California and Iowa, more than 90 percent. South-East Asia has the largest mangrove ecosystem in the world, but it is rapidly being clear-cut by commercial logging operations, especially in Malaysia, Indonesia, and Bangladesh.

Like many other environmental problems, but especially so, the world's wetlands need international cooperation for their protection. The significance of wetlands, and the consequences of their loss, cannot be contained by national borders. The migrating birds of Western Europe, for example, depend on wetlands in northern Africa for refuge on their journey; the drainage of wetlands in Louisiana threatens the fisheries of the entire Gulf of Mexico.

Since 1971, some measure of international wetlands protection has been offered through the Convention on Wetlands of International Significance (also known as the 'Ramsar Convention'), but this convention is ratified by only 43 countries, and while 77,000 square miles of wetlands have fallen under its statutory protection, this represents less than 2 percent of the world's wetland areas.

Sources for map and commentary:
Chapman, V.J., *Ecosystems of the World and Wet Coastal Ecosystems*, New York: Elsevier, 1977; Conservation Foundation, *State of the Environment: A View Toward the Nineties*, Washington DC, 1987; Houck, Oliver, 'America's Mad Dash to the Sea,' *The Amicus Journal*, Summer 1988; 'Mangroves: An Ecosystem in Danger', *Ambio*, vol. 9, 1980; Saenger, P., E.J. Hegerl & J.D.S. Davie, *Global Status of Mangrove Ecosystems*, IUCN (Cambridge) Commission on Ecology Papers no. 3, 1983.

15 WASTE

International information on levels of municipal and household waste generation, most experts agree, is little more than an educated guess, although some general points are clear: worldwide generation of waste appears to amount to about one billion tonnes per year, and is growing rapidly; the biggest waste producers are rich countries and, within poor countries, rich people. Affluence produces effluence, literally.

There is also a significant difference in the kind of garbage thrown away by rich and poor: in rich countries, glass, paper, metals and other durables constitute a larger share of garbage than in poor countries, where more of the waste is organic material. This difference has important implications for waste disposal: organic garbage can be composted, which is the safest method of waste disposal, whereas other materials are less easily disposed of. Plastics are the least disposable of any garbage, yet the proportion of plastics in garbage is increasing dramatically. In 1987, the USA alone produced 26 billion kgs. of plastics; by volume, plastics now represent about 30 percent of the US waste stream.

At the beginning of the 1990s, most garbage in most places in the world is dumped in landfills. But cities and countries everywhere are running out of landfill space. Given the landfill crisis, attention is turning to incineration (optimistically called, by those in the industry, 'waste to energy' disposal). Incineration is a high-cost means of waste treatment that generates air pollution and leaves a highly toxic residue that itself then needs disposal. A number of countries, including Belgium, France, West Germany, and the UK, burn some of their waste at sea on incinerator ships, although the international community is moving closer to banning this highly polluting and very dangerous practice. Other countries simply dump their municipal garbage at sea. Although this practice, too, is being phased out, we will live with its legacy for years to come, in the form of increasing epidemics of marine mammal illnesses, and increasing incidents, some of which are already recorded in Western Europe and the USA, of deep-sea trash washing up on beaches.

Recycling, which is a cheap, revenue-generating, and low-technology solution, is slowly gaining favour in most industrialized countries, but still faces heavy competition from incineration. In poor countries, recycling has always ranked high in waste reduction, and whole communities of people in most Third World cities support themselves as dump-scavengers and recyclers.

The waste crisis has sparked interest among rich countries in the possibilities of an international waste trade. With no room in their own backyards, states and industries in the developed world are scanning the globe for countries poor enough to want to accept municipal trash for (usually low) dumping fees. (See Map 27 for a parallel trade in toxic waste). In 1989, an American waste disposal firm closed a deal to dump thousands of tonnes of US household waste on a Pacific atoll in the Marshall Islands. It is a great irony that one of the selling points for the deal was that the garbage could provide landfill for the low-lying Marshalls, which would otherwise be vulnerable to flooding if sea levels rise as a result of global warming.

Sources for map and commentary:
Cointreau, Sandra, *Solid Waste Recycling: Case Studies in Developing Countries*, Washington DC: World Bank, 1987; Levenson, Howard, Office of Technology Assessment, US Congress, personal communication; Organization for Economic Cooperation and Development (OECD), *OECD Environmental Data Compendium, 1989*, Paris: OECD, 1989; World Resources Institute/Institute for Environment and Development, *World Resources 1988-89*, New York: Basic Books, 1988; World Action on Recycling Materials for Energy from Rubbish (WARMER), Tunbridge Wells, personal communication; Worldwatch Institute, *State of the World: 1989*, New York: Norton, 1989.

16 SEWAGE

In heavily populated areas, especially the rapidly-growing urban conglomerations of poorer countries, sewage pollution is a primary cause of the worldwide trend in deterioration of water quality. In developing countries, most municipal sewage receives little or no treatment, and separation of drinking water from sewage disposal is inadequate; sewage contamination of drinking water is a major factor in the high rates of infant mortality and morbidity recorded in many poorer countries, much of which is caused by gastro-intestinal infections.

In older industrial areas, municipal water and sewage facilities, many built in the last century, are overwhelmed by population pressures, and a decline in investment in sewage disposal in the late 1970s and early 1980s has resulted in the downgrading of river water quality in most urban areas. Until recently, sewage in most industrialized countries was dumped directly into the nearest body of water: for example, until the mid-1960s, American and Canadian cities pumped untreated sewage directly into the Great Lakes, contributing to its reputation as one of the most polluted bodies of water in the world. In most cases, it was not municipal farsightedness but agitation by independent environmental groups that led to pressure for sewage treatment and an end to direct dumping. Efforts to clean up urban waterways, now underway in many industrialized cities, are estimated to cost billions of dollars.

The accuracy of statistics that measure the provision of sanitation facilities is widely suspect. International compendiums typically provide dramatically different figures: for example, while the World Health Organization reports adequate sanitation provisions for 40 percent of Uganda's urban population in the mid-1980s, an independent development agency reports only 20 percent coverage. In some cases, statistics may be manipulated to serve political agendas that might be undercut by reports of low levels of public health services provision. These statistical variations, often dramatic or confusing, suggest the difficulties inherent in attempts to monitor public and environmental wellbeing on a global basis.

Sources for map and commentary:
Organization for Economic Cooperation and Development (OECD), *OECD Environmental Data Compendium, 1989*, Paris: OECD, 1989; United Nations Environment Programme (UNEP), *Environmental Data Report*, 2nd ed. 1989-90, Oxford: Blackwell, 1989; World Health Organization (WHO), *World Health Statistics Annual*, 1986; World Resources Institute/Institute for Environment and Development, *World Resources 1988-89*, New York: Basic Books, 1988.

17 AIR QUALITY

This map depicts urban levels of three major air pollutants - sulphur dioxides, nitrogen oxides, and particulates ('dust and dirt'). Ninety percent of anthropogenic sulphur dioxide emissions are from the burning of certain fossil fuels; nitrogen oxides come primarily from car and truck emissions; and particulates from both cars and fossil fuel combustion. All three contribute to a range of respiratory illnesses, including bronchitis, pneumonia, asthma and lung cancer, although the health effects vary with the intensity and duration of exposures. Both sulphur dioxide and nitrogen oxides are major components of acid rain and nitrogen oxides are also a minor greenhouse gas.

Measuring air quality is a precise science that must, by its nature, be carried out under imprecise and unpredictable conditions. Standardized international measurements are only available for a scattering of cities, and drawing comparisons from city data is difficult because of wide variations in place, time and the frequency of air monitoring samples. Sometimes, indeed, monitoring is only carried out at sites where there is a known severe problem. This leads to a bias towards sampling cities with higher concentrations of pollutants. Despite these caveats, though, it is still possible to make certain generalizations about air quality problems in the world's cities.

The air in most cities is in fact a chemical soup of dozens of pollutants. There are a number of others beyond the three pollutants illustrated on the map. Carbon monoxide, largely the product of car emissions, is one of the mostly widely distributed urban air pollutants, but there is no available international data set measuring its levels. Similarly, ozone, also known as 'photochemical smog', is a serious air quality concern in many cities, but again there are no worldwide statistics. Los Angeles is still the city with the greatest ozone problem in the world. Ozone was originally believed to be a problem peculiar to Los Angeles, because of its special topography and high density of cars, but photochemical smog is now known to occur in and downwind of most major cities in the world. Dublin also has a serious smog problem - but in this case, the smog is from the 350,000 tonnes of coal burned in the city's households each year.

Most urban air pollutant levels are increasing, largely because of increases in the numbers of cars and trucks (see Map 31). There are some broad estimates of the costs involved in reducing air pollutant levels: that to cut sulphur dioxide emissions in half in the USA would cost an estimated US$5 billion a year; to halve sulphur emissions in the EEC would cost from US$5 billion to US$7 billion per year. It is even more difficult to quantify the real costs of existing air pollution, but one EEC estimate gives US$13 billion for loss of fisheries, crops, forests, and health.

Sources for map and commentary:
Global Environment Monitoring System (GEMS), *Assessment of Urban Air Quality*, Nairobi: UNEP/WHO, 1988; 'Monitoring the Global Environment: An Assessment of Urban Air Quality,' Special Issue report, *Environment*, October 1989; Organization for Economic Cooperation and Development (OECD), *OECD Environmental Data Compendium*, 1989, Paris: OECD, 1989; World Health Organization (WHO), *Air Quality Guidelines for Europe*, Copenhagen: 1987; World Commission on Environment and Development, *Our Common Future*, Oxford: Oxford University Press, 1987; World Resources Institute/International Institute for Environment and Development, *World Resources 1988-89*, New York: Basic Books, 1988.

18 ENERGY BUDGETS

Energy production and consumption are closely monitored, both by government agencies and private industry, and statistics on international use are widely available, if sometimes highly technical. As always, information on commercially-traded commodities, such as oil or coal, is much more reliable than is information on fuels of local reliance, such as firewood and small-scale hydropower. Firewood, the largest biomass energy source, and other biomass fuels together provide approximately 14 percent of the world's energy.

The USA is by far the largest consumer of commercial energy, but a number of other countries in Western Europe have comparable per capita consumption levels. Current global energy consumption rates are increasing on an average of 3 percent per year, the largest share of which is in the industrialized world. Overall, it would appear that the pace of growth in consumption is somewhat lessening, but consumption rates often fluctuate considerably over the short term.

Energy consumption provides a measure of our impact on ecological systems. The production, extraction and use of coal, oil, gas, and nuclear power, which power much of the world, all have destructive environmental consequences (see Maps 17, 22 and 23). The renewable energy sources most widely used, hydro power and firewood, are more environmentally benign, but are not unproblematic (see Maps 20 and 21). Firewood is in increasingly short supply, and even though it provides the sole energy source for many 'third world' households, it may not continue to be a viable energy source far into the next century.

Given this array of poor choices, energy conservation is crucial to the long-term viability of the planet. After the oil crisis of 1973, systematic efforts were made throughout the industrial world to increase energy efficiency, with impressive results: in the USA, cars will now travel 29 percent farther on a gallon of gasoline than they did in 1973; Denmark was able to reduce its total use of direct fuels by 20 percent between 1976 and 1980. The reasons to reduce energy consumption are now different but they are even more compelling.

Some would argue that conservation is necessary but not sufficient, and that we must radically alter our industries and economies to incorporate large-scale use of solar, geothermal, and wind energy sources, which hold the promise of being both renewable and non-polluting. Others see them as introducing environmental problems of their own. As yet, however, these technologies have received little support or interest, nor have they made much of a dent in traditional energy budgets.

Sources for map and commentary:
Conservation Foundation, *State of the Environment: A View Toward the Nineties*, Washington DC, 1987; Goldsmith, Edward & Nicholas Hildyard, eds., *The Earth Report*, London: Mitchell Beazley, 1988; United Nations, *Energy Statistics Yearbook 1987*, 1989; US Department of Energy, Energy Information Administration (EIA), *International Energy Annual, 1988*, Washington DC: DOE, 1989.

19 THE NUCLEAR FAMILY

Electricity generation by nuclear power began in 1955 with a small reactor in the Soviet Union. Since then, the growth of the industry has been dramatic. There are currently 421 operating nuclear power plants in the world, and another 144 reactors planned or under construction. In addition to the 32 countries with operating power plants, another 27 have some form of nuclear facility. Though the industry is widely perceived to have fundamental problems, there are actually few signs of its demise.

The USA is the world's largest exporter of nuclear technology, followed by the USSR; it is not coincidental that these two countries are also the world's largest military powers, and the world's largest military exporters. Military nuclear agendas and commercial nuclear programmes are closely intertwined. Any nuclear power reactor can be used to produce plutonium for bombs. Some, like 'fast breeder' reactors, are designed specifically to produce plutonium; others produce it as a by-product; many reactors are used for both weapons and energy production. Whether governments choose to acknowledge the interconnectedness of the military and civilian uses of nuclear programmes depends largely on domestic politics and international approbation.

The link between civilian nuclear power and military nuclear capability is made through the nuclear fuel cycle. To be able to make a nuclear weapon, a country has to have access to all five stages of the nuclear cycle, from uranium production through fuel reprocessing. Thus, South Korea, **111** which is technically striving for independence in its nuclear power capabilities, has no enrichment or reprocessing facilities; on the other hand, Argentina, which does not produce much power through nuclear facilities, has completed the fuel cycle. The fuel cycle also brings countries into the nuclear family even if they are not producing nuclear energy: Iraq, for example, has a nuclear fuel fabrication plant and Gabon mines uranium.

Spokespeople for the nuclear industry are quick to point out that nuclear power plants emit none of the greenhouse gases produced by oil and coal. This, however, is a dubious virtue, given that the operation of nuclear plants represents an ongoing public safety and health threat, and given the large volumes of highly toxic waste generated by the nuclear industry. Over the next 30 years, more than 25,000 cubic metres of high level waste will be generated by the top ten nuclear countries alone. There is literally no place to safely dispose of radioactive waste, and there are fewer and fewer places to store it. Until 1983, Belgium, the Netherlands, Switzerland and the UK dumped low-level radioactive waste at sea. This practice was halted by an international moratorium under the London Dumping Convention, an agreement consistently opposed by the British government. With the ban on ocean disposal, there is also no place, any more, to dispose of low-level radioactive waste. Most nuclear states are building up huge stockpiles of nuclear waste that no one wants and no one knows how to get rid of.

The nuclear accident at Chernobyl in 1986 provided powerful evidence that nuclear disasters transcend state borders and undermined support for nuclear power throughout Europe. Polls showed that opposition to nuclear power in the UK jumped frpm 70% before Chernobyl, to 85% after the accident; in Austria, opposition increased from 50% to 75%; in France, from 40% to 50%. In a national referendum in 1987, Italians voted to shut down permanently their nuclear power plants, although as of early 1990, it appeared that one plant was still operating.

Sources for map and commentary:
Commission of the European Communities, *The State of the Environment in the European Community, 1986*, Brussels, 1988; International Atomic Energy Agency *Annual Report for 1988*, Vienna: IAEA, 1989; 'The World's Nuclear Fuel Cycle Facilities', *Nuclear Engineering International*, December 1987; 'World List of Nuclear Power Plants', *Nuclear News*, August 1989.

20 HYDRO POWER

In 1950, there were about 5000 large dams in the world; thirty years later, by the early 1980s, there were 35,000 large dams, 53 percent of which were in China. The frenzy of dam construction appears to have now slowed worldwide, and in China in particular - in large measure because of environmental concerns.

Hydro power (which accounts for about 6 percent of worldwide energy use) is in itself a relatively cheap, renewable and non-polluting energy source. However the dams built to produce it present a number of serious environmental problems. The reservoirs for large dams are created by flooding vast swathes of land, often inundating valuable forest area, wildlife habitat or wetlands - often entire ecological communities. In warm climates, reservoirs can pose health hazards, increasing the incidence of waterborne disease, such as schistosomiasis, and 'river blindness'. The quality of water from dams often deteriorates, due to a build-up of salts and chemicals, posing threate to irrigated crops and drinking water supplies.

The construction of a large dam is a huge undertaking. Many of the largest dam projects in the world are funded with assistance from international aid agencies, most notably the World Bank. The twin Sadar Sarovar and Narmada Sagar dam projects in India, for example, estimated to displace 70,000 people by submerging 438 villages, was started with a US$450 million loan from the World Bank; it has become one of their most controversial projects. If the World Bank honours its recent commitment to phase out support for environmentally destructive projects, many of the largest dam projects now planned around the world will be halted. Local environmental pressure is also halting work on some large dams (see Map 37). Plans for a major dam on the Danube at Nagymaros, Hungary, were scrapped in 1988. In 1990, French environmentalists are fighting proposals for four large dams on the Loire, France's longest river. They have had some success. Two dams have already been cancelled and the battle continues.

The definition of 'large' dams, as agreed by international engineering associations, includes dams from 15 metres in height to 'mega' dams over 200 metres high. Statistics on hydro power, and on dams, are readily available from 'pro-dam' sources, mainly detailed records of dam characteristics and construction, and from 'anti-dam' groups who monitor the environmental and social costs of dams.

Sources for map and commentary:
International Commission on Large Dams, *World Register of Dams*, Paris: 1984; International Rivers Network, *World Rivers Review*, July/August, 1989; Mermel, T.W., 'The World's Major Dams and Hydro Plants', *International Water Power and Dam Construction*, vol. 41, 1989; Parcels, S. & T.B. Stoel, 'The Large Dam Controversy', Unpublished draft for the World Bank, 1989; Probe International, 'Water Projects With World Bank Involvement', *World Rivers Review*, September/October 1989; Simons, M., 'Brazil Wants its Dams, But at What Cost?' *New York Times*, March 12, 1989.

21 FIREWOOD

Firewood is a local, self-reliant energy source. Since most statistical information is collected on a national or even international level, information on the supply and use of firewood is partial, inconsistent, and often based on very small samples. Official reports also invariably underestimate rural fuelwood use.

Within a single country there is often considerable variation in firewood provision – variations lost in national summaries. Information on the amount of time spent collecting firewood – always women's work -- must be treated with a good deal of caution. The data we provide on this subject is based on a small sample of communities and cannot be assumed to represent a general or national average.

Nonetheless, the general situation in relation to firewood is clear. Many people in many 'third world' countries rely entirely on firewood for energy, cooking, and warmth. About 70 - 95 percent of all firewood is used for rural domestic purposes, and the remainder is used in cottage industries and small-scale commercial enterprises. Almost everywhere in the world, firewood is becoming scarce, and the effort to collect it is taking a larger and larger share of the working day. Many women in rural areas report that they are now spending so much time collecting firewood that they have little time for other activities, such as growing and cooking food. With less time to prepare food, simpler diets are being adopted – in the Sahel, for example, many women have shifted from cooking millet to rice. Nutrition suffers even before a full firewood crisis arises.

The firewood shortage has been precipitated by a number of converging factors. Everywhere in the world, deforestation for commercial logging or agriculture is the primary cause of local firewood scarcity. In addition, though, urban demand is putting pressure on rural resources. Firewood for Delhi, for example, comes from Madhya Pradesh, 700 kilometres away, and requires clear-felling of 6 hectares of forest a day. Whereas fuelwood for rural uses is often twigs and fallen branches, commercial users will cut down an entire tree – representing the difference between sustainable tree use and deforestation.

Sources for map and commentary:
Agarwal, Bina, *Cold Hearths and Barren Slopes: The Fuelwood Crisis in the Third World*, Riverdale, MD: Riverdale Press, 1986; Food and Agriculture Organization (FAO), *Map of the Fuelwood Situation in the Developing Countries*, Rome: FAO, 1981; FAO, *Yearbook of Forest Products, 1982, 1984*; Richard Hosier, University of Pennsylvania, personal communication; Smil, V. & W.E. Knowland, eds., *Energy in the Developing World: The Real Energy Crisis*, Oxford: Oxford University Press, 1980; United Nations Environment Programme (UNEP), *Environmental Data Report*, 2nd ed. 1989-90, Oxford: Blackwell, 1989; United Nations Statistical Office, *Energy Statistics Yearbook 1985,1987* and *Energy Statistics Yearbook 1986*, 1988.

22 OIL POLLUTION

The economy – and lifestyle – of the industrialized world is entirely dependent on oil. Oil is big business, both for private and public enterprise and state revenue: Exxon, the world's largest oil company made a net profit of US$5 billion in 1987; 55 states depend essentially on a single product for their export income – for 24 of them, that product is oil and gas. The oil industry seems larger than life; so, unfortunately, is the pollution it causes. The extraction and transport of oil are among the highest-risk industrial activities, causing long-lasting ecological disruption and widespread pollution.

More than 6 million metric tonnes of oil are released into the world's oceans each year; roughly 1 tonne is spilled for every 1000 tonnes extracted. Marine tanker accidents are not the largest source of oil pollution, but they are often the most dangerous, releasing large amounts of oil in a short period of time, often close to shore. The threat posed by single acute oil spills has increased over the past twenty years, as oil companies have switched to larger and larger vessels in an effort to cut costs. Today's supertankers average twice the size and carry ten times the amount of oil as tankers that worked the oil routes twenty years ago.

The chronic release of oil from maritime shipping operations represents an even greater environmental problem than acute spills. On oil supply routes, for example, oil tankers routinely flush out their tanks with sea water on the return trip, in so doing releasing thousands of tonnes of oil annually. Oily trails around the world mark the major shipping routes.

When accidental maritime oil spills do happen, as they do on an average of three a day worldwide, legal and financial responsibility for the ensuing pollution is often difficult to assign. Many oil companies register their ships under 'flags of convenience' in such countries as Liberia, the Bahamas, Cyprus or Panama. These 'open registries' allow companies to run ships with lower-cost, non-union crews, to evade certain corporation taxes, and to avoid some national and international regulations. Registries of convenience also obscure the trail of corporate responsibility for accidents and spills.

In mapping the distribution of oil spills, our original intention had been to allocate responsibility to individual oil companies. This proved to be impossible. Despite the volumes of data kept by tanker associations, oil associations, and maritime associations, no one, it appears, monitors oil spills by company – or, if such records do exist, they are not open to public scrutiny. This gap reinforces the sense that oil accidents 'just happen', somehow without human agency. It is even difficult to establish, in the first place, the occurrence of spills. Unless they are spectacular accidents, most domestic media coverage ignores spills that occur elsewhere in the world, and most small spills in domestic waters are not newsworthy. The data used here is a compilation from dozens of sources, and yet we know the information to be still incomplete.

A primary argument against increased use of renewable energy sources, such as solar power, is its cost. It is argued that conventional sources of energy, such as oil, are much cheaper. However, the true cost of oil is mostly hidden. The real tally of the cost of our oil dependency should include the cost of environmental damage and clean-ups. It should also include expenditure on military forces used to 'secure' oil supply routes and oil production regions; oil company subsidies and tax allowances, and the cost of the trade deficits that most oil-dependent nations now face.

Sources for map and commentary:
Couper, A., ed., *The Times Atlas of the Oceans*, London: Times Books; New York: Van Nostrand, 1983; International Tanker Owners' Pollution Federation (ITOPF), 'Spills over 5000 barrels,' mimeograph document; Levy, Eric, 'Oil Pollution in the World's Oceans,' *Ambio*, vol. 13, no. 4, 1984; McKenzie, Arthur, *Guide to the Selection of Tankers*, New York: Tanker Advisory Center, 1989; other Tanker Advisory Center publications; National Academy of Sciences (NAS), *Oil in the Sea: Inputs, Fates and Effects*, Washington DC: NAS, 1985; Organization for Economic Cooperation and Development (OECD), *OECD Environmental Data Compendium, 1989*, Paris: OECD, 1989; Starr, Gary & Susan Bryer Starr, 'The True Cost of Oil,' *Earth Island Journal*, Summer 1988.

23 FOSSIL FUEL POLLUTION

Fossil fuels are oil, gas and coal. Most industrial economies are entirely fossil-fuel dependent. The price of this dependency is severe pollution and, looming, the possible onset of global warming. The burning of fossil fuels produces a number of pollutants, including sulphur dioxide, nitrogen oxides, particulates (dirt and dust) and carbon monoxide.

The most serious fossil fuel pollutant is carbon dioxide. This is not in itself toxic, but in conjunction with other gases is the primary greenhouse gas (see Map 1). Most carbon dioxide in the atmosphere stems from burning fossil fuels. Worldwide releases from carbon dioxide from fossil fuel combustion currently amount to 22 billion tonnes per year. Other human activities that produce carbon dioxide include tropical forest destruction (see Map 5), cement production and natural gas burn-off flares. Smaller quantities are produced from natural sources, such as volcanoes and animal respiration.

Most fossil fuel pollution comes from a handful of rich, industrial countries. Canada and the USA produce, between them, 33 million tonnes of carbon dioxide a year. Carbon dioxide pollution is a pollution of privilege; but as is often the case in environmental concerns, the consequences and costs are borne globally.

Attempts to curb the problem are being thwarted by the very countries most responsible. In autumn 1989, 65 countries sponsored an international resolution to cut carbon dioxide emissions in industrial countries. It was defeated by objections from Japan, the USA and the USSR. The US and Japanese governments urged postponing action for 'further study', while USSR and other East European representatives opposed pollution limits claiming they could not afford them. In the countries creating most pollution, the adoption of carbon dioxide limits would probably require more fuel-efficient cars, more fuel-efficient electricity and extensive tree planting.

The other pollutants included on this map, sulphur dioxide and nitrogen oxides, are both major contributors to acid rain (see Map 24). Again, the primary sources of these pollutants are in the industrialized world and acid rain is a particular problem for Western Europe. Most West European governments, the UK being a notable exception, have joined together to form a '30% Club'. This commits them to reducing sulphur dioxide emissions by 30 percent by 1993 – not just by cutting back on fossil fuel use, but by persuading industry to devise less polluting techniques.

Sources for map and commentary:
Eliasson, Anton, et al, *Estimates of Airborne Transboundary Transport of Sulphur and Nitrogen over Europe*, Norwegian Meteorological Institute, 1988; Global Environment Monitoring System (GEMS), *Assessment of Urban Air Quality*, Nairobi: UNEP/WHO, 1988; Goldsmith, Edward & Nicholas Hildyard, eds., *The Earth Report*, London: Mitchell Beazley, 1988; Montgomery, Paul, 'US, Japan and Soviets Prevent Accord to Limit Carbon Dioxide', *New York Times*, November 8, 1989; Oak Ridge National Laboratory, *Estimates of CO2 Emissions from Fossil Fuel Burning and Cement Manufacturing*, Environmental Sciences Division Publication no. 3176, 1989.

24 ACID RAIN

The term 'acid rain' includes both the wet deposition of acidic sleet, snow, fog or rain and the dry deposition of nitrates and sulphates as they settle. Since the industrial revolution the chemical balance of the atmosphere has changed and acidification is being increased. This is mainly through the release of a very large additional burden of sulphur dioxide and nitrogen oxides from burning fossil fuels, notably coal in power stations and oil in motor vehicles (see Maps 23 and 31).

The acidity of a solution is measured in terms of 'pH'; pH1 is strongly acid, pH14 strongly alkaline, and pH7 is neutral. Weak acids such as lemon juice or vinegar have a pH of 3 to 4. Natural rainfall is usually slightly acid and has a pH of 6 to 7. On the map, 'high' acid rain pollution corresponds to an average rainfall pH of 4.9 to 5.1, 'very high' to pH4.5 – 4.7, and 'extremely high' to pH4.2 – 4.3.

Acid rain spares little. It erodes railway tracks; it corrodes the masonry of historic buildings and public statues. It causes natural ecosystems to crumble, destroying forests, crops and fish life in **115** lakes. To date, acid rain is doing the greatest damage in Europe. In southern Norway, 80 percent of the lakes and streams are technically 'dead'; 64 percent of UK forest is showing signs of acid rain damage; in central Switzerland, more than 40 per cent of conifer forest is dead or severely damaged.

Acid rain is, by definition, a trans-national problem – it shows no regard for state boundaries. Acid deposition has been a contentious issue between Canada and the USA for years; most acid deposition in Scandinavia is from other countries, coming downwind from Europe's major industrial centres.

Solutions to the acid rain problem are energy conservation, a shift to less polluting forms of transport, and the fitting of special equipment to reduce acid gas emissions: 'flue gas desulphurization' at power stations and three-way catalytic converters on motor vehicles. In 1985, 19 countries agreed to reduce sulphur emissions by a further 30 percent, but scientists estimate that the reduction must be very much higher if the lakes and forests of Scandinavia are to be saved from further damage.

Sources for map and commentary:

Goldsmith, Edward & Nicholas Hildyard, eds., *The Earth Report*, London, Mitchell Beazley, 1988; McCormick, John, *Acid Earth*, London: Earthscan, 1989; United Nations Environment Programme (UNEP), *Forest Damage and Air Pollution*, 1988; World Resources Institute/International Institute for Environment and Development, *World Resources 1988-89*, New York: Basic Books, 1988.

25 INDUSTRIAL WASTE

Industrial activity generates millions of tonnes of waste annually, much of it extremely toxic. Nowhere in the world are there effective constraints or regulations to control this practice, and the decision-makers in industry seem largely unwilling to police themselves. Industry likes to conduct 'business as usual' and the typical response is to fight tooth and nail against the demands of clean air legislation, chemical regulation, water quality standards, automobile emission standards, workplace health regulations, and any other legislation designed to protect our environment. In state-controlled economies, where the state and industry are one, there are even fewer opportunities for independent monitoring or control of industrial pollution.

Some 350 million tonnes of hazardous industrial wastes are generated worldwide each year, about 90 percent of which comes from industrial countries. The US Environmental Protection Agency has identified a top-priority national list of over 5000 hazardous sites ('Superfund sites') left behind by industry, which, together will cost billions of dollars to clean up. As the curtain of secrecy lifts from Eastern Europe, tales of dead rivers, unbreathable air, villages abandoned because of extreme pollution, and other industrial horrors are being revealed daily.

There are no good ways of getting rid of hazardous waste. In the UK, 82 percent of hazardous industrial waste is dumped in landfills (see commentary to Map 6). At the beginning of the 1990s, the UK was still dumping 8 percent of its hazardous waste at sea. Protests from other EEC countries and from environmentalists mean that this practice is to be phased out. In the USA, incineration is increasingly favoured by industry, but this creates other forms of pollution.

Detailed information on the extent, nature, and origin of industrial pollution is patchy: there are powerful forces at work to ensure that much of it remains unidentified. The 'blighted zones' identified on this map represent the most extreme cases of industrial pollution; literally thousands more zones of local blight scar every industrial country.

Sources for map and commentary:

Bellini, James, *High Tech Holocaust*, San Francisco: Sierra Club Books, 1986; National Wildlife Federation, *The Toxic 500*, 1987, New York: NWF, 1989; press reports; United Nations Environment Programme (UNEP), *Environmental Data Report*, 2nd ed. 1989-90, Oxford: Blackwell, 1989.

26 DEADLY INDUSTRY

This map identifies the major accidents of the past twenty years that have occurred during the manufacture and transport of hazardous industrial products. It does not include accidents that occur in their end-use and application. There are several biases in the data. More is known about accidents in the 'developed' world and, by definition, the focus is on acute events rather than chronic releases (see Map 25). Then there is a general absence of accidents involving weapons factories, military research laboratories and other military installations because of a dearth of information.

Industrial accidents involving the release of hazardous substances are a daily occurrence. But the record-keeping of these releases is entirely inadequate. Only large and spectacular accidents, such as the explosion of gas tanks in Mexico City in 1984, attract international attention, and most smaller-scale accidents do not even attract notice from a national press. In the USA, several quasi-official groups attempt to monitor accidents – the Acute Hazardous Events Data Base (AHE/DB) and the National Response Center of the Department of Transportation among them – but few other countries have systems for tracking industrial accidents.

What we do know about industrial accidents is that they are no accident. They have come to be accepted adjuncts of an industrial culture that, using materials of ever-greater toxicity, is extending its reach around the world.

Sources for map and commentary:
Gittus, J.H., et al, *The Chernobyl Accident and Its Consequences,* London: UKAEA, 1988; Kasperson, Roger, et al, *Corporate Management of Health and Safety Hazards: A Comparison of Current Practice,* Boulder, CO: Westview Press, 1988; Kleindorfer, Paul & Howard Kunreuther, *Insuring and Managing Hazardous Risks: From Seveso to Bhopal and Beyond,* Berlin: Springer, 1987; Lagadec, Patrick, *Major Technological Risk: An Assessment of Industrial Disasters,* Oxford: Pergamon Press, 1982; Organization for Economic Cooperation and Development (OECD), *OECD Environmental Data Compendium, 1989,* Paris: OECD, 1989; press reports, 1988-90; United Nations Environment Programme (UNEP), *Environmental Data Report* 2nd ed. 1989-90, Oxford: Blackwell, 1989.

27 TOXIC TRADE

Industrialized countries produce over 300 million tonnes of hazardous waste each year. These wastes pose a threat to human and animal health and any number of towns and ecosystems around the world have been poisoned by improper disposal of toxins. The dilemma of toxic waste is that there is really no 'proper' or safe way to dispose of it; if dumped in landfills or storage pits, poisons almost inevitably leak out into drinking water and residential areas; if burned, it leaves toxic residue; if dumped at sea (now banned by most countries), it threatens marine mammal and fish life. In lieu of solving the waste problem, many governments and private firms have chosen simply to ship the problem away – out of sight, out of mind. Poor countries are targeted to be recipients of the rich world's industrial waste – they need the revenue that dumping contracts can provide, and they often do not have the resources independently to assess the potential health and pollution effects of the waste they accept.

Between 1986 and 1988, over three million tonnes of waste were known to be shipped from rich to poor countries. This figure represents the tip of the iceberg; much of the trade is illicit or illegal, and its actual extent is difficult to determine. The international waste disposal business is a shady one: in many cases toxic waste shipments have been illegally dumped on the shores of poor countries; in other cases, governments of Third World countries are deliberately misled about the nature of proposed shipments.

The toxic trade has roused Third World anger – such countries have no interest in being the rich world's dumping ground – and almost 40 countries, mostly in the Third World, have now banned toxic waste shipments.

In addition to the Third World dumping trade, many industrialized countries exchange waste, although the flow of trade still reflects power imbalances between exporters and importers. Canada **117** receives a large share of US toxic waste; East Germany, Hungary and the UK are major repositories for European waste; China accepts waste from the USA and from Europe. This trade may be more consensual than Third World dumping, but it is no less dangerous.

Sources for map and commentary:
Organization for Economic Cooperation and Development (OECD), *OECD Environmental Data Compendium, 1989*, Paris: OECD, 1989; Vallette, Jim, *The International Trade in Wastes: A Greenpeace Inventory*, Washington DC: Greenpeace, 1989.

28 HOLES IN THE SKY

A thin layer of ozone, high in the stratosphere, acts as a shield against harmful solar ultraviolet radiation. A reduction in ozone allows more ultraviolet radiation to reach the earth. This is likely to result in long-term increases in skin cancers and damage to the human immune system. It will also lead to disruption of agriculture, reductions in crop yields, and possible large-scale alterations in a number of ecosystems. Ozone depletion is already a considerable and growing problem; the rate of destruction is so extensive that 'holes' in the ozone layer – varying in size, depending on the time of year - have formed over and close to the polar regions.

Since the mid-1970s, scientists have known that chlorine from synthetic compounds called CFCs (chlorofluorocarbons), and to a lesser extent a number of other chemicals, destroy ozone in the stratosphere. CFCs are entirely artificial compounds – they were developed in the 1930s and since then have been the basis of very powerful and profitable chemical industries. For many years, these industries, as well as governments in CFC-producing countries, dismissed theories of ozone depletion and resisted increasingly urgent calls from the scientific community for curbs in the production of CFCs and other ozone-depleting chemicals. The connection is now accepted, however, and the weight of opinion has finally shifted.

Agreement on the most appropriate and rapid solution is more difficult to achieve. The Montreal Protocol represents a first step. It was first signed in 1987 by 24 countries; by the end of 1989 it had been signed by 48 countries and agreed to in principle by another nine non-ratifying states (although ratification usually follows quickly, it cannot be assumed). Ratifying countries in the industrialized world agree to freeze CFC production at 1986 levels, with allowances for continued growth in the Third World for ten more years. While this agreement is a major step forward, environmentalists consider that it falls far short of what is necessary. The Montreal Protocol does not stop CFC production and ozone depletion – it merely slows down the pace of acceleration.

The US Environmental Protection Agency estimates that each 2.5 percent increase in CFCs causes an additional one million skin cancers, which will result in 20,000 additional deaths over the lifetime of the existing US population. Since the 1960s, CFC production has increased at an average 5 percent per year. In the mid-1980s, worldwide CFC production was over one million tonnes per year. The USA is the world's largest producer (and consumer).; in Europe, the UK is the largest producer. The uses of CFCs vary; in Europe aerosols account for 37 percent of CFC use; in the USA, 39 percent of all CFCs are used in motor vehicle air-conditioning and refrigeration. Globally, aerosols are the largest source of CFC emissions.

Sources for map and commentary:
US Environmental Protection Agency, *Effects of Changes in Stratospheric Ozone and Global Climate*, Washington DC: 1986; Hammit, J., *Product Uses and Market Trends for Potential Ozone-Depleting Substances, 1985-2000*, Santa Monica, CA: Rand Corporation, 1986; ozone hole graphics from the NASA Ozone Trends Panel, 1989, courtesy John Gille, National Center for Atmospheric Research, Boulder, CO; Shea, Cynthia Pollock, 'Protecting the Ozone Layer' in Worldwatch Institute, *State of the World: 1989*, New York: Norton, 1989; United Nations Environment Programme, *The Ozone Layer*, Nairobi: UNEP, 1987.

29 WAR-WASTED LANDS

Despite an encouraging number of ceasefires, peace talks and arms control talks, warfare was still widespread in 1989, affecting nearly a fifth of the world's states. Since 1945, most warfare has taken place in the 'Third World', and most wars raging in the late 1980s were civil wars. There is no universally accepted definition of war. This map includes wars defined politically – armed unrest involving quasi-military structures, as in Northern Ireland – as well as wars defined by high death tolls.

Wars are acutely destructive to the environment. Often, as in Vietnam in the 1960s and 1970s and in Central America in the 1980s, the natural environment is itself a military target. In Nicaragua, the US-backed contras made targets not only of the environment, but of environmentalists. Between 1982 and 1989, contras are known to have killed 30 environmental workers and kidnapped 70 more, forcing the closure of several national parks and halting several water pollution control projects. The largest contiguous area of unspoiled rainforest in Central America is to be found in Guatemala; by the late 1980s, one-third of Guatemala had been sprayed with defoliants as part of the government's anti-insurgency campaign, and also as part of the US-initiated 'war against drugs'. In the Vietnam war, US herbicide teams dumped millions of gallons of herbicides and defoliants (chief among them the now-notorious Agent Orange) over half the forests in the country; more than 50 per cent of the coastal mangrove swamps in the south were destroyed by bombs, defoliants, napalm, and bulldozers, and most will never regenerate. In the Iran-Iraq war, major wetlands were destroyed by bombing, chemical exposure and deliberately-set fires.

Even when the environment is not a designated target, it inevitably suffers. Agricultural areas become unworkable as villages are destroyed, farming families are killed or forced to flee, crops and livestock are seized or destroyed. And as in Afghanistan, large tracts of land are strewn with mines or with unexploded bombs and shells.

The environmental effects of war are rarely documented systematically, and thus this map can only suggest the nature and extent of the problem.

Sources for map and commentary:
Environmental Project on Central America (EPOCA), Green Paper Series, 1985 to present, Earth Island Institute, California; Stockholm International Peace Research Institute (SIPRI), *World Armaments and Disarmament: SIPRI Yearbook*, 1987, 1988, 1989, Oxford, OUP; Sivard, Ruth Leger, *World Military and Social Expenditures 1989*, Washington DC: World Priorities Inc., 1989; Wallensteen, Peter, ed., *States in Armed Conflict 1988*, Uppsala University: Department of Peace and Conflict Research, Report no. 30, July 1989.

30 NUCLEAR BLIGHT

In the late 1980s, the world nuclear weapons stockpile was approximately 50,000 warheads, about 16,000 of which are deployed at sea. Only five states acknowledge possession of nuclear weapons, although Israel and South Africa are also widely believed to be nuclear states. A number of other states allied with the USA and the USSR have nuclear weapons based on their territories, willingly or not. The development, production, storage, transport and deployment of nuclear weapons requires an extensive support network of bases and facilities, each one of which poses a permanent environmental hazard.

Recent revelations in the US media underscore the chronic nuclear pollution and heavy environmental damage caused by military nuclear facilities. Leaks of radioactive materials from the Rocky Flats facility threaten the drinking water supply of Denver; the Fernald materials production plant in Ohio has released more than 300 tonnes of uranium dust into the surrounding air and water; the agricultural land around Hanford, Washington, shows high levels of radiation.

In addition to chronic nuclear pollution, there are dozens of accidents each year involving military storage and transport of nuclear weapons or material. In this map, we focus on the most serious, known, nuclear accidents at sea – in which either nuclear reactors and/or nuclear weapons have been lost. Many more accidents may have happened but been kept secret. There are currently **119**

about 50 nuclear weapons and seven nuclear reactors known to be resting on the ocean floor.

There would be even more nuclear material dumped at sea without strong international opposition. As recently as 1989, the British government was proposing to dump its nuclear submarines on the sea-bed once they are obsolete. The US government has officially rejected sea dumping as an option for getting rid of old nuclear military equipment, but the current administration is known to be reconsidering this possibility. France, the UK and the USA oppose a permanent ban on dumping although non-nuclear states are almost universally in favour.

The Pacific Islands have borne the greatest brunt of nuclear might and nuclear blight. Long used as nuclear testing grounds by France, the UK and the USA, whole island chains in the Pacific have been 'vaporised', while others are so radioactive as to be uninhabitable for thousands of years. The populations of many Pacific territories have become nuclear refugees. The Pacific nuclear-free movement, largely a women's movement, is now one of the most active grassroots anti-nuclear campaigns in the world.

Sources for map and commentary:
Arkin, William & Joshua Handler, *Naval Accidents 1945-1988*, Neptune Papers no. 3, Washington DC: Greenpeace/ Institute for Policy Studies, 1989; Arkin, William & Richard W. Fieldhouse, *Nuclear Battlefields*, Cambridge, MA: Ballinger, 1985; Kidron, Michael & Dan Smith, *The War Atlas*, London: Pan; New York: Simon & Schuster, 1983; Radioactive Waste Campaign, *Deadly Defense*, New York: RWC, 1988; Stockholm International Peace Research Institute (SIPRI), *World Armaments and Disarmament: SIPRI Yearbook*, 1968/69, 1974, 1975, 1985, 1989, London: Taylor & Francis; Oxford, OUP.

31 AUTO CULTURE

In much of the world, most people still rely on bicycles, public buses and trains, animal-drawn carts and their own feet for transport. But the global trend is clearly towards automobile-centred transport.

The costs are high. Private automobiles clog the streets of most world cities bringing urban services to a standstill; expanding road networks gobble up scarce space and degrade agricultural land, especially in developments on the outskirts of cities; the fuel demands of private transport deplete fossil fuel resources. Worldwide, at least a third of an average city's land is devoted to roads, parking, and other car-related infrastructures. One of the most serious consequences is that cars play a prominent role in generating virtually all the major air pollutants (see Maps 17 and 23). In OECD countries, for example, 75% of carbon monoxide emissions, 48% of nitrogen oxides and 13% of atmospheric particulates (dirt and dust) are produced by cars and trucks; in the USA, motor vehicles are responsible for about 70% of total carbon monoxide emissions, 40% of nitrogen oxides, and 20% of particulates. Lead pollution, a serious health problem in most cities, is largely attributable to motor vehicle emissions.

The growing emphasis on private car ownership usually comes at the cost of public transport. Governments often assign priority to private motorized travel in traffic planning, budget decisions, and urban design. As the emphasis on private transport increases, funding and planning for public transit declines proportionally. This has the effect of further marginalizing people on the economic and social fringes of society – the poor, the elderly and, especially, women of all classes and ages. Women everywhere in the world are much more dependent on public transport since they are less likely than men to own private automobiles. A survey taken in London in the mid-1980s, for example, established that men drive cars twice as often as women, only a third of British women have a driving licence and 70 percent do not have a car. Automobile ownership both reflects and reinforces social inequities. The social structures that promote and reward car ownership will need to change if we are to be serious about environmental protection.

The global growth in private automobile production and ownership is fueled by automobile manufacturers who have a vested interest in the continuous expansion of their markets. As rich countries become saturated with automobiles, the Third World beckons as the next promising market. Car culture already has a firm grip on the large cities of Mexico, Brazil, and India, and is making strong inroads almost everywhere else in the developing world.

This map includes only information on cars. In some countries, such as China, the number of trucks and buses exceeds the number of cars and these, too, must be considered as major contributors to urban clog and air pollution.

Sources for map and commentary:
Mick Hamer, 'Men have Wheels While Women have Feet', *New Statesman*, May 24, 1985; Stan Luger, 'Stagnant Politics, Dirty Air', *In These Times*, December 13-19, 1989; Motor Vehicle Manufacturers' Association, *World Motor Vehicle Data: 1989*, New York, 1989; Michael Renner, 'Rethinking Transportation', Worldwatch Institute, *State of the World: 1989*, New York: Norton, 1989.

32 TOURIST TRAPS

By the mid-1980s, the global tourism business employed more people than the global oil industry; by the end of this century, tourism will be the world's largest economic activity. Tourism is increasingly the largest single revenue source for poor countries, and most governments actively promote the development of tourist trade. Without tourism, the economies of many of the Caribbean nations, for example, would collapse, as would the economies of many Mediterranean coastal towns.

The ratio of tourists to residents is a guide to the strain which tourists place on their host environment. High rates of tourism can overwhelm water supply, sewage treatment and municipal refuse facilities. St. Maartens, a small island in the Netherlands Antilles, receives the equivalent of its own population in tourists every 10 days. Tourist traffic and its related infrastructure can be a serious threat to fragile environments and often wipes out natural habitats entirely. To serve the tourist trade, many governments have embarked on ambitious programmes of road, airport, and hotel construction, often without serious regard for the environmental consequences of these developments. The wildlife trade, which threatens many rare species (see Map 33) often flourishes with tourism: rare and exotic birds and animals are hunted to sell to tourists – either alive, or as ivory trinkets, tortoise-shell combs, stuffed-toys made of animal fur, brilliantly-coloured corals, and the like.

The figures on tourism, collected by the World Tourism Office, monitor visitor arrivals. These include arrivals by guest workers or multiple border crossings by the same person or family. A few countries, Hungary for example, thereby appear to receive more tourists than they do in reality.

Another niche in the market has been spotted and tourism is now being heralded as a potential force for environmental good. The development of national parks and game reserves in many countries, most noticeably in Costa Rica and a number of African countries, is spurred on by the lure of tourist revenue. But present evidence suggests at best an uneasy coexistence of 'eco-tourism' and environmental protection.

Sources for map and commentary:
Commission of the European Communities, *Quality of Bathing Water: 1987*, Luxembourg: 1989; Enloe, Cynthia, *Bananas, Beaches and Bases: Making Feminist Sense of International Politics*, London: Pandora Press, 1989; World Tourism Organization, *Yearbook of Tourist Statistics: 1988*, Geneva, 1989.

33 WILD TRADE

Statistics on the extent of the traffic in wildlife are, at best, unreliable. An international body, the Convention on International Trade in Endangered Species (CITES) monitors the trade, and imposes restraints on the trade in certain animals designated for protection. Currently, trade is officially banned for about 675 species (identified as 'Appendix I' species), and regulated for at least 27,000 more ('Appendix II' species). CITES figures are the best available, but are known to be underestimates. Although CITES does make estimates of the scope of illegal trade, it is almost **121** impossible to verify these figures; the information presented on this map refers to both legal and illegal trade, although it must be assumed that most illegal trade is not reported. The representation of CITES statistics on this map includes the secondary trade in processed animal products and the re-export traffic – thus some countries are identified as both major exporters and importers of the same animals.

For the two inset maps, on the top 12 ivory and wildcat fur trading countries, we have used total trade by volume, both import and export, for the mid-1980s.

The trade in wildlife is almost entirely a luxury trade: the demand comes from rich countries, the supply from poor countries. The trade rests on a triad of commodification, greed, and desperation. To the brokers in wealthier countries, animals are mere commodities, exotic 'things' to be traded as expensive baubles. It is difficult to interpret the selling of wildlife, often pursued to the brink of extinction, as anything other than a sign of desperation. The closest parallel is perhaps prostitution; in extreme poverty, all the sellers have left to sell is themselves. In many poor countries, wildlife is what there is left to sell.

The traffic in animals is also a male enterprise. It is mostly men who poach the animals in poor countries; or men who organize expensive hunting safaris from rich countries; it is men who act as brokers of the international trade; and it is men who purchase the animals and their by-products: rhinos for aphrodisiacs, for example; or elephants, for ivory-handled daggers.

The traffic in wildlife is big business. The annual legal trade is worth about US$5 billion annually, and illegal trade is estimated at another $2 billion. In one year alone, 1985, the known trade included 43,000 primates, almost 9 million reptile skins, and 720 tonnes of ivory. While animal counts are imprecise, it appears that the trade in wildlife, dead and alive, has reduced the number of snow leopards left in the world to about 500, reduced tiger populations in India from about 40,000 to less than 3000, precipitated a decline in world rhino numbers of 70 percent since 1970, and reduced the herds of African elephants to about 625,000 out of an original population of over two million. It is widely thought that the only hope for preventing extinction of the elephant may be an absolute ban on ivory production and consumption. In 1989, the CITES countries (see pages 90-97) voted to do just this, but some African countries, such as South Africa and Zimbabwe, claim that their managed herds offer greater protection. CITES has been unable to persuade Hong Kong from selling its vast stockpile of imported ivory, though this is known to encourage illegal poaching.

Sources for map and commentary:
IUCN Conservation Monitoring Centre, *Red List of Threatened Animals*, Cambridge: IUCN, 1988; United Nations Environment Programme (UNEP), *UNEP News*, December 1989; World Resources Institute/International Institute for Environment and Development, *World Resources 1988-89*, New York: Basic Books, 1988.

34 HUNTING

Comparatively few human populations meet their food and clothing needs by hunting any more; and among those that do, their subsistence hunting leaves relatively little mark on the ecosystem. This map focuses on recreational hunting, whether for 'sport' alone or to supply the luxury trade in wildlife (see Map 33).

The primary source of information on animal extinctions is the IUCN's *Red List of Threatened Animals* which is published every two years. Species extinctions are often caused by a number of stress factors, such as habitat loss combined with over-hunting. The species identified on this map are those for whom the primary threat appears to be luxury or sport hunting.

Recreational hunters often defend their sport by saying that it is a celebration of nature, but the record speaks otherwise. Around the world recreational hunting has pushed hundreds of species to the brink of extinction and, over recent history, hundreds more have been hunted out of existence. Some species, shown on the map as endangered, are virtually extinct already. For example, by 1990 over much of eastern Africa, there were only 55 remaining Java rhino, about 800 Kemp's ridley turtles and fewer than a thousand black rhinos.

Despite repeated calls by the international community for an end to whaling, five countries still hunt whales. Whaling is conducted by these countries under the pretext of 'gathering scientific information' on whale populations, but in fact whaling is still the basis of a small but thriving commercial enterprise.

Hunting is big business. Apart from the fantastic sums of money sometimes paid for trophy hunting of rare animals (or merely parts, such as gorilla's paw or rhino horn), the ordinary business of hunting generates considerable revenues. Hunters in the USA spend around US$16 billion a year on fees and equipment. In poorer countries, hunting is often a major source of external revenue earned from rich foreigners. In 1983, Zimbabwe for example, earned US$6.2 million in fees and licenses from hunting for sport. Yet given the damage they do and the revenue that is generated, surprisingly few people hunt: in the USA, 7%; in Western Europe, 2%, in the USSR, 1%. Everywhere hunting is primarily a men's activity, and may often be construed as a rite of passage into manhood.

Sources for map and commentary:
Burton, J.A. and B. Pearson, *Rare Mammals of the World*, New York: Collins, 1987; Fitter, R., *Wildlife for Man*, New York: Collins, 1986; Inskipp, T. & S. Wells, *International Trade in Wildlife*, London: Earthscan, 1979; IUCN Conservation Monitoring Centre, *Red List of Threatened Animals*, Cambridge: IUCN, 1988; Organization for Economic Cooperation and Development (OECD), *OECD Environmental Data Compendium, 1989*, Paris: OECD, 1989; World Resources Institute/International Institute for Environment and Development, *World Resources 1988-89*, New York: Basic Books, 1988.

35 THE TIMBER TRADE

The tropical forests of the world are rich in highly-prized, 'luxury' woods such as teak, mahogany and balsa, among dozens of others. Extracting this timber is a high- profit, fast-moving big business. Most commercial logging operations in the tropics are monopolized by a handful of multinational timber conglomerates based in France, West Germany, the UK and Japan. They follow in the footsteps of the colonial traders in tropical timber, and in many ways replicate patterns of colonial exploitation – marked by utter disregard for the environment of Third World countries, perceived primarily as suppliers of raw materials to meet the 'refined tastes of the 'developed' world.

Uncontrolled logging has left many tropical forests in critical condition. Deforestation has already become a major problem in some of the earliest centres of commercial logging. Thailand, for example, once a major exporter, became a net importer of wood in 1980. Nigeria is virtually logged out. Many forests in Southeast Asia and West Africa are in a critical condition. Of the 56 timber-exporting countries, 23 are now net importers.

Timber is a valuable commodity of wide commercial interest - so valuable that the statistics are considered to be privileged commercial intelligence and are not widely available to the public. Independent environmental groups, particularly the UK-based Friends of the Earth, appear to have the most comprehensive accessible data on both the timber trade and the state of tropical forests. The International Tropical Timber Organization, formed in 1987 and based in Japan, is currently the only international body with oversight of the timber trade, but this organization's close ties with industry make suspect its effectiveness as an independent monitoring agency.

Sources for map and commentary:
Myers, Norman & Richard A. Houghton, *Deforestation Rates in Tropical Forests, and Their Climatic Implications*, London: Friends of the Earth, 1989; Nectoux, F. & Nigel Dudley, *The Hardwood Story*, London: Friends of the Earth / WWF 1987.

36 PAYING THE PRICE

In the course of an ordinary day, which many us live in a fog of pollution, we encounter a barrage of harmful chemicals, most of them invisible and many of them unknown – in our workplaces, on the streets, in our food and in our homes. In much of the world, air and water quality are, at best, suspect. That environmental derangement on this scale brings with it severe health repercussions cannot be doubted. As environmental damage escalates, so, too, will problems of health. However, **123** specific and definitive links are often difficult to establish. As yet, only a few environmental problems are known to have identifiable health consequences - such as the depletion of the ozone layer (see Map 28) causing rising rates of skin cancer. However, the hazards of urban living are clearly illustrated by increased lead levels found in the blood of most inhabitants of most large cities.

Exposure to carcinogenic chemicals and low-level radiation are the two principal causes of cancer and there has been little action to limit either. It is thus not surprising that, worldwide, cancer rates are rising – the World Health Organization reports six milion new cancer cases each year.

Lung cancer, the most common cancer in the developed world, is known to be caused by a number of factors, including occupational hazards and chemical exposures, but its primary cause is smoking. Tobacco fuels an extremely profitable and commercially aggressive industry. As rates of smoking decline in many industrial countries, tobacco firms are now exerting heavy pressure on Third World consumers, especially women, to take up smoking. Trends in lung cancer follow trends in smoking habits. As the lifestyles and environmental problems of the industrial world spread around the globe, so does lung cancer.

The increasing use of tobacco in many parts of the world has important implications for the environment. As a lucrative cash crop, tobacco is replacing traditional and subsistence crops in many Third World countries. Tobacco depletes soil nutrients and encourages heavy use of fertilizers and herbicides. Further, the World Health Organization estimates that one tree in six is cut for use in curing tobacco.

Health statistics are fairly complete for the industrial world, and unavailable or inconsistently available for most of the rest of the world. In any case, internationally available health statistics are not presented in conjunction with any discussion or evidence of contributing causes. Environmental causality, which must be superimposed on the raw health data, is usually best illustrated on the level of very specific and very closely-analysed case studies, not on a national or global level.

Sources for map and commentary:
Vahter, Marie, *Assessment of Human Exposure to Lead and Cadmium through Biological Monitoring*, Stockholm: UNEP, 1982; World Health Organization (WHO), *World Health Statistics*, 1988.

37 A CLASH OF INTERESTS

The global environment is a commons, shared by all countries. Increasingly, countries also share in its deterioration. Most pollution and degradation of the environment cannot be contained within national borders. Acid rain from the rest of Europe is devastating Sweden's forests and lakes; water diversion plans for irrigation in Ethiopia will diminish Egypt's water supplies; deforestation in Nepal, China, Bhutan and India has increased the frequency and extent of destructive flooding in Bangladesh. An estimated 40 percent of the world's population depends, for drinking water, irrigation or hydro power, on major river systems shared by two or more countries; in some cases, waterways are shared by as many as five or more countries.

In most cases, this share-out of environmental problems is inevitable. Occasionally, though, governments and private industries take active measures to 'ship' their pollution abroad. They may build smokestacks tall enough to disperse airborne pollutants, for example, or export unwanted and hazardous waste to poorer countries.

Increasingly, relations between governments are mediated by environmental issues. As certain resources are diminished and transboundary pollution increases, the environment is moving closer to the centre of international relations. Similarly, as individuals become more informed about environmental issues, national governments are judged on the efficacy of their environmental programmes. 'Green consciousness' poses a challenge to traditional notions of national sovereignty and international diplomacy, and to established forms of governance.

Sources for map and commentary:
Press reports; Renner, Michael. *National Security: The Economic and Environmental Dimensions*, Washington DC: Worldwatch Paper no. 89, May 1989.

INDEX